2016国际动物考古协会理事会
暨全球发展与中国视角动物考古学术研讨会论文集

动物考古

第 3 辑

河南省文物考古研究院　编

文物出版社
北京·2019

图书在版编目（CIP）数据

动物考古. 第3辑／河南省文物考古研究院编. ——北京：文物出版社，2019.12
ISBN 978-7-5010-6419-9

Ⅰ.①动… Ⅱ.①河… Ⅲ.①古动物学-文集 Ⅳ.①Q915-53

中国版本图书馆CIP数据核字(2019)第256345号

审图号：GS（2020）1806号

Henan Provincial Institute of Cultural Heritage and Archaeology, ed. 2019. Zooarchaeology vol. 3. Proceedings of Zhengzhou 2016 International Committee Meeting of ICAZ & "Archaeozoology: Global Developments and Chinese Perspectives". Beijing: Cultural Relics Press.

动物考古（第3辑）

编　　者：河南省文物考古研究院

责任编辑：王　媛
封面设计：程星涛
责任印制：张道奇
责任校对：陈　婧
英文校对：温成浩

出版发行：文物出版社
地　　址：北京市东直门内北小街2号楼
邮　　编：100007
网　　址：http://www.wenwu.com
邮　　箱：web@wenwu.com
经　　销：新华书店
印　　刷：北京京都六环印刷厂
开　　本：787mm×1092mm　1/16
印　　张：12.25
版　　次：2019年12月第1版
印　　次：2019年12月第1次印刷
书　　号：ISBN 978-7-5010-6419-9
定　　价：139.00元

本书版权独家所有，非经授权，不得复制翻印

Proceedings of Zhengzhou 2016 International Committee Meeting of ICAZ
& "Archaeozoology: Global Developments and Chinese Perspectives"

ZOOARCHAEOLOGY

Volume 3

Edited by
Henan Provincial Institute of Cultural Heritage and Archaeology

Cultural Relics Press
Beijing · 2019

前　言

2016 年 10 月 13 日至 16 日，由河南省文物局主办，河南省文物考古研究院承办的"2016 国际动物考古协会理事会暨全球发展与中国视角动物考古学术研讨会"在郑州召开。这是国际动物考古协会首次在亚洲举办理事会。来自美国、加拿大、墨西哥、阿根廷、法国、德国、瑞典、土耳其、南非、澳大利亚、日本、印度和中国共 13 个国家的 18 位国际动物考古协会理事出席了理事会。本次理事会就协会工作进行了认真商讨，对未来两年的重要学术活动进行了安排，并就加强国际动物考古学者与中国同行的交流合作，提高中国动物考古学的国际化水平，吸纳更多来自世界不同文化、语言背景的新成员等议题达成了广泛共识。同时，来自美国哈佛大学、法国国家自然历史博物馆、北京大学、吉林大学、山东大学、中国科学院古脊椎与古人类研究所、中国社会科学院考古研究所、河南省文物考古研究院、陕西省考古研究院、湖北省文物考古研究所等 34 个海内外高校和科研院所的 44 位动物考古学者，围绕"全球发展与中国视角动物考古"这一主题，对考古遗址出土动物遗存基本信息、古代动物的饲养与屠宰模式、骨制品生产及其蕴含的人类社会意识等议题展开了深入的探讨，对动物考古资料数字化、大数据分析和数据库共享等未来的研究发展方向进行了展望。

本文集收录的 12 篇论文是从会后提交的论文中挑选出来的，其中英文 7 篇、中文 5 篇，是继 2014 年出版的《动物考古·第 2 辑》之后的第三本动物考古专辑。文集采用中文和英文两种语言。所收录的中文论文附有加长的英文摘要，以便国外学者参考使用。以下对文集收录的 12 篇论文作扼要介绍。

Alice M. Choyke 的《从山川去平原，从平原回山川——亚美尼亚 Godedzor 遗址铜石并用时代晚期骨器研究》对亚美尼亚南部一处季节性使用的铜石并用时代晚期的高地遗址 Godedzor 出土的骨器进行了研究。通过宏观形态观察和微痕分析，作者认为该遗址出土的骨器主要用于纺织和编织活动。同时，这些季节性返回高地的小规模的游牧族群可能与西部较远的制造业传统以及其他成熟的生产领域建立了联系，这种联系甚至可以到达伊朗北部地区。

Erika Gál 的《匈牙利东南部 Szegvár – Szőlőkalja 遗址 7 世纪阿瓦尔墓地的双管鸟骨笛研究》对匈牙利东南部阿瓦尔早期（7 世纪）墓地 Szegvár – Szőlőkalja 出土的一支双管骨笛进行了研究。该骨笛出土于一座成年女性墓葬，由两根骨管构成，每根骨管上有五个乐孔，第五个乐孔均已残损。经鉴定，两根骨管由同一只鹤的左右两侧跗跖骨制成。鹤曾是匈牙利大平原上常见的动物，它的腿骨是制作骨笛的理想材料。对比同地域其他遗址出土的骨笛，作者认为此类骨笛较为罕见。这种双管鸟骨笛所吹奏的旋律可能在 7 世纪末有着特殊的含义。

Hans Christian Küchelmann 的《德国东部青铜时代中期到铁器时代早期的骨器组合研究》对德国柏林附近的凯姆尼茨遗址出土的五件青铜时代中期到铁器时代早期的骨器进行了研究。研究表明，五件骨器均制作简单，原料来自野兔、家猪、山羊以及大型哺乳动物（牛或马鹿的体型）。根据对骨器表面微痕的观察及模拟实验，作者认为其中四件尖状骨器可能有多重用途，另外一件骨器的用途有待进一步探讨。

Manuel Altamirano García 的《维护，继承和记忆——西班牙格拉纳达 Las Peñas de los Gitanos 的 Los Castillejos 铜石并用遗址的 V 形穿孔象牙纽扣研究》，采用实物资料与文献相结合的方法，对制作于公元前 3000 年并经过长期流传使用的 9 个 V 形穿孔象牙纽扣进行了研究。这些纽扣由贵重的象牙原料精心制作而成，是财富和地位的象征。通过裸眼和扫描电镜的观察，这些纽扣有长期使用而磨损的光滑表面，并经多次修复，可见曾被世代相传。观察所得的信息以及考古学和民族学的背景资料，是解释这批纽扣制造、使用、管理和废弃等方面不可或缺的材料。

Pam J. Crabtree 的《乌克兰东部 Razdolnoe 遗址青铜时代晚期动物遗存研究》，对 2010 年夏季乌克兰东部的 Razdolnoe 遗址发掘的动物遗存进行了鉴定和研究。遗址出土的动物种属包括牛、绵羊、猪、马、狗等家养动物，以及欧洲野牛、马鹿、狍子等野生动物。虽然样本量较少，不足以对遗址出土动物的屠宰模式和年龄结构等问题进行探讨，但为研究当地青铜时代晚期动物群和动物资源的获取方式提供了基础资料。

Shaw Badenhorst 和 Douglas Ross 的《加拿大不列颠哥伦比亚省狮子岛 19 世纪末 20 世纪初华人社区遗址出土的动物遗存研究》对加拿大西海岸弗雷泽河下游狮子岛上 19 世纪末 20 世纪初华人社区出土的 1218 件动物遗存进行了研究。这些动物遗存包括猪、牛、海狸、麝鼠、鲨鱼等，以猪为主，表明当时在遗址生活的华人以猪肉为主要肉食来源，这与中国人传统的肉食结构基本一致。同时，这一研究结果与 19 世纪晚期到 20 世纪早期北美西部大多数的华人居住遗址情况相似。

László Bartosiewicz、Gertrud Haidvogl 和 Clive Bonsall 在《多瑙河沿岸鲟鱼捕捞的动物考古和历史资料》一文中，对史前和中世纪多瑙河沿岸鲟鱼捕捞的动物考古和历史资料进行了梳理和分析。研究表明，鲟鱼在多瑙河的灭绝受到了多种人类活动的影响，包括水路运输增长、过度捕捞和水坝建设等。作者强调，在处理传统的鲟鱼捕捞、灭绝以及重新引入多瑙河这类生态和经济问题时，需要从多学科综合研究的角度出发。

陈全家、王雅艺和王春雪的《吉林大安后套木嘎遗址出土贝类遗存研究》一文，在对吉林大安后套木嘎遗址出土的贝类遗存种属鉴定的基础上，对不同贝类的时空分布和贝壳产品的加工工艺、流程及功能进行了分析。研究认为，遗址中贝类种属及比例未随时间发展发生明显变化，周围自然环境比较稳定；不同遗迹类型中贝类种属比例不同，先民会有意识地根据需求选择相应的贝类使用；随着时间的推移，贝类制品加工过程中产生的废料减少，加工工艺日趋成熟；先民已熟练掌握打制、磨制、锯割、剔刮和穿孔等技术，并将这些技术运用到工具及装饰品的加工制作中，能够进行大量的标准化制作。这些信息有助于了解后套木嘎居民的工具制作流程、手工业发展水平及生活场景，进而揭露居民生产生活的整体面貌。

侯亮亮、张国硕、戴玲玲和侯彦峰的《辉县孙村遗址殷墟文化时期动物骨骼的稳定同位素分析》对辉县孙村遗址出土的殷墟文化时期的猪、狗、牛、羊、鹿等26例标本进行了碳、氮稳定同位素分析，并与同时期临近的殷墟遗址已发表的人、家猪和家犬的稳定同位素数据进行了对比。结果显示，家猪和家犬的食物结构不能成为重建殷商文化时期先民生业经济的替代性指标，但可以作为大部分先民生业经济间接的、笼统的指示参考；随着社会复杂化程度的加深，外来及新的农作物的推广和种植，以及家畜专业化饲喂和管理的强化，将家猪和家犬的食物结构作为重建先民生业经济参考时需要更加慎重。

李凡的《黄牛与牦牛骨骼形态的对比观察》一文，通过观察和测量河南省文物考古研究院动物考古实验室、中国社会科学院考古研究所科技考古中心动物考古实验室收藏的至少16个黄牛现生骨骼标本、10个牦牛现生骨骼标本，对黄牛和牦牛的前颌骨、下颌骨、寰椎、枢椎、掌骨、跖骨、第一指/趾节骨、第二指/趾节骨等部位的骨骼形态差异进行了描述。该研究有助于在今后的动物考古实践中进一步开展黄牛与牦牛的骨骼形态比较研究。

王运辅在《考古遗址出土啮齿目遗存的采集与鉴定方法》一文中简要介绍了在考古遗址中采集啮齿目遗存的基本方法，并根据动物考古学研究的实际需要，以褐家鼠、黑家鼠、黄胸鼠、小家鼠等家栖鼠类以及多种与人类生活密切相关的野生啮齿动物为实例，讨论了啮齿目鉴定工作中需要注意的事项，如应重点观察的若干解剖特征点，包括头骨的基本形态，各种脊、突、孔、窝等细微的结构，以及基本的颊齿类型等。

余翀的《大数据及其在动物考古学中的应用——中国新石器时代至青铜时代早期生业模式的特征》一文，介绍了大数据的相关概念及其在西亚和欧洲的动物考古学研究中的成功应用案例，并利用地理信息系统软件对中国199处新石器时代至青铜时代早期遗址出土的动物遗存中各类动物的相对比例进行了初步考察。结果表明，自然环境差异、家养动物的出现和传播等因素与各类动物相对比例的区域性和历时性变化有着密不可分的关系。

从文集收录的12篇论文可以看出，对某一考古遗址出土的动物遗存和骨器的研究仍是中外动物考古学者关注的重点，而对动物骨骼形态鉴定、数据统计、科技分析的方法的探索与应用已经成为近年来中国动物考古发展的新风向。本文集为我们了解中外动物考古的最新研究动态提供了信息，也希望能为中国动物考古学的发展助力赋能。我们要特别感谢百忙中为文集赐稿的每位作者，没有他（她）们的支持与合作，就不可能出版这本文集。

马萧林

2019年12月

Preface

Archaeology is the study of ancient material culture and as such, many of its problems are strongly regional, relevant to past cultures whose geographical distribution was often relatively limited. The study of archaeological animal bones, on the other hand, has a potential to be far more globalized. Even if there can be substantial zoogeographical differences between major areas, the animal skeleton is a far more universal structure than products of human creativity. Thanks to this 'common language' international cooperation should be natural to our field of research. On the other hand, the local treatment and use of animals and especially the way animal products were used was always an integral part of ancient material culture. This way animal osteologists, sharing a universal method, have important regional information to be shared with colleagues, archaeologists and zooarchaeologists alike. The increasing international scope of this series is a reflection of this situation. Volume 2 of Zooarchaeology was the proceedings of the 9th meeting the Worked Bone Research Group of the International Council for Archaeozoology (ICAZ) held in Zhengzhou in 2013. ICAZ, our world – wide organization also had its 2016 committee meeting in Zhengzhou. In the meantime, a new generation of scholars from universities across China began regularly attending international meetings joining a network of zooarchaeologists. Increasing mutual engagement by our Chinese colleagues and their counterparts around the world exposes the wealth of academic information on both sides that has been isolated from one another by history and language. Beyond mere information, however, the potential for active learning from and cooperating with each other has tremendously increased. Authors in this volume have demonstrated through their work the importance of such exchanges. Special thanks are due to the series editors of Zooarchaeology who have recognized the importance of such links and have supported them through their continuing efforts.

Alice M. Choyke and László Bartosiewicz

December 2019

目 录

从山川去平原，从平原回山川

  亚美尼亚 Godedzor 遗址铜石并用时代晚期骨器研究 …………… Alice M. Choyke / 1

匈牙利东南部 Szegvár – Szőlőkalja 遗址 7 世纪阿瓦尔墓地的双管鸟骨笛研究

  ………………………………………………………………………………… Erika Gál / 14

德国东部青铜时代中期到铁器时代早期的骨器组合研究 ………………………………

  ………………………………………………… Hans Christian Küchelmann / 21

维护，继承和记忆

  西班牙格拉纳达 Las Peñas de los Gitanos 的 Los Castillejos

  铜石并用遗址的 V 形穿孔象牙纽扣研究 ………… Manuel Altamirano García / 29

乌克兰东部 Razdolnoe 遗址青铜时代晚期动物遗存研究 ………… Pam J. Crabtree / 41

加拿大不列颠哥伦比亚省狮子岛 19 世纪末 20 世纪初华人社区遗址出土的动物遗存研究

  ……………………………………………… Shaw Badenhorst，Douglas Ross / 46

多瑙河沿岸鲟鱼捕捞的动物考古和历史资料

  ………………… László Bartosiewicz，Gertrud Haidvogl，Clive Bonsall / 59

吉林大安后套木嘎遗址出土贝类遗存研究 ……………… 陈全家　王雅艺　王春雪 / 72

辉县孙村遗址殷墟文化时期动物骨骼的稳定同位素分析

  ………………………………………… 侯亮亮　张国硕　戴玲玲　侯彦峰 / 91

黄牛与牦牛骨骼形态的对比观察 ………………………………………… 李　凡 / 106

考古遗址出土啮齿目遗存的采集与鉴定方法 ………………………… 王运辅 / 113

大数据及其在动物考古学中的应用

  中国新石器时代至青铜时代早期生业模式的特征 ………………… 余　翀 / 127

后记 ………………………………………………………………………………… 142

Contents

From the Mountains to the Plains and Back Again
   Late Chalcolithic Worked Osseous Material from Godedzor, Armenia
   .... Alice M. Choyke / 1
Bird Bone Double Pipe from the 7th Century Avar Cemetery of Szegvár-Szőlőkalja (South-East Hungary) .... Erika Gál / 14
Middle Bronze to Early Iron Age Bone Tools from Eastern Germany
   .... Hans Christian Küchelmann / 21
Maintenance, Inheritance and Memory
   V-perforated Ivory Buttons from the Los Castillejos Chalcolithic Site in Las Peñas de los Gitanos (Granada, Spain) .... Manuel Altamirano García / 29
Bronze Age Faunal Remains from Late Bronze Age Features at Razdolnoe, Eastern Ukraine .... Pam J. Crabtree / 41
The Fauna from Lion Island, a Late Nineteenth and Early Twentieth Century Chinese Community in British Columbia, Canada .... Shaw Badenhorst, Douglas Ross / 46
Archaeozoological and Historical Data on Sturgeon Fishing along the Danube
   .... László Bartosiewicz, Gertrud Haidvogl, Clive Bonsall / 59
The Study of Shell Remains from the Houtaomuga Site in Da'an, Jilin
   .... Quanjia Chen, Yayi Wang, Chunxue Wang / 72
Stable Isotopic Analysis of Animal Bones from the Suncun Site, Huixian, Henan, China during the Yinxu Culture (ca. 1250 – 1046 BC)
   .... Liangliang Hou, Guoshuo Zhang, Lingling Dai, Yanfeng Hou / 91
Differences in Osteological Morphology between Cattle (*Bos taurus*) and Yak (*Bos grunniens*) .... Fan Li / 106
Methods of Collection and Identification of Rodent Remains from Archaeological Sites .... Yunfu Wang / 113
The Application of Metadata in Zooarchaeology
   Evaluating Subsistence Patterns in Neolithic and Early Bronze Age China by Using Published Mammal Records .... Chong Yu / 127
Afterword 142

From the Mountains to the Plains and Back Again

Late Chalcolithic Worked Osseous Material from Godedzor, Armenia

Alice M. Choyke

Medieval Studies Department, Central European University, Hungary

Abstract: The over two hundred late Chalcolithic and Early Bronze Age bone, antler and tusk tools from the upland seasonal occupation at Godezor in south Armenia represent a new data point on the scholarly bone tool map of this period in the southern Caucasus region. Although there is little contemporary comparative material, some useful information can be gleaned from this material. Based on form and macro wear (10 × to 20 ×), it appears that many of these objects may have been used in spinning and weaving activities. It is also clear that the small pastoralist groups that returned to the upland settlement seasonally were connected to broad horizons of manufacturing traditions further to the west as well as probably to other established spheres of production elsewhere. The wear on these objects shows some came from outside the settlement from more distant regions and more sophisticated workshop environments, perhaps located in the urban centers in Northern Iran. There are currently no published assemblages of worked bone from Late Chalcolithic urban centers in northern Iran. Other objects from Godezor are less well thought out, more *ad hoc*, and can be connected to daily activities during the local seasonal occupation. The assemblage on the whole reflects exactly the kinds of objects, both sophisticated and simple, connected widely and also intensely local that one would expect to result from a seasonal occupation by transhumant pastoralists.

Keywords: bone industry, Late Chalcolithic, Armenia, pastoralism

INTRODUCTION

Research on later prehistoric worked osseous materials from the Chalcolithic onwards in the Caucasus is at its beginnings. The material from the site of Godezor, located in south Armenia near the small town of Sisian in Syunik Province, represents a single point in the history of such materials during the Late Chalcolithic and beginning of the Early Bonze Age Kura-Axes Culture in the region. Thus, there is virtually no comparative material with which to contextualize the assemblage. Nevertheless, some typological and use wear patterns are apparent. Some of the objects were made and used at this small pastoralist settlement in the highlands of southern Armenia whereas other objects appear to come from more sophisticated workshop environments, perhaps be brought onto the site as the population, or part of the population at least, moved over the landscape from lowland plains in what is today northern Iran to highland areas in what is today southern Armenia.

The site of Godedzor is located at 1800 m a. s. l. on a high volcanic terrace above the Vorotan river gorge in a highland area overlooking a small plain, today dominated by the village of Angeghakot appropriate for a late spring to early fall occupation by transhumant pastoralists (Avetisyan et al. 2009; Chataigner 2011: 69; Chataigner 2016: 1 – 4; Kalataryan et al. 2008; Kalataryan et al. 2010; Palumbi et al. 2007; Palumbi and Chataigner 2014). The Vorotan River links the steppes of Azerbaijan (Agdam region) to Nakhichivan and the Araxas valley down into north-west Iran. The area is covered by heavy snow from November to March. There are important obsidian resources in the area. It has been hypothesized that the choice of this spot for transhumant pastoralists was linked not only for graze for sheep and cattle herds but also to the trade of obsidian (Chataigner et al. 2010: 251).

Excavations began at the site in 2003 and ended in 2013 (https://missioncaucase. hypotheses. org/godezor-armenie, accessed October 29, 2018). The responsible institutions for excavations were the Institute of Archaeology and Ethnography in Yerevan in Armenia with field work led by Irena Kalataryan and the UMR 5133 Archéorient, Lyon in France with field work led by Giulio Palumbi. The archaeologists in charge were Pavel Avetisyan (Institute of Archaeology and Ethnography in Yerevan, Armenia) and Christine Chataigner (UMR 5133 Archéorient in Lyon, France). The over-whelming bulk of the features and findings comes from the Late Chalcolithic. The site itself comprises a series of circular stone structures (Plate Ⅰ. Figure 1) in a rocky landscape abutting still higher cliffs. It is now used as a stone quarry, an activity that will eventually endanger the site itself. The setting of the site is quite dramatic as it is placed around a large central natural upright stone with a view to mountains in the east (Plate Ⅰ. Figure 2). It was used as a transhumant camp by pastoralists at the end of the Late Chalcolithic in the mid-fourth millennium to the very end of the fourth millennium or for around 500 years. There is some evidence of an earlier Chalcolithic occupation based on scattered ceramics, a slightly later Early Bronze Age occupation as well as ephemeral traces of Iron Age occupation on this site (Avetisyan et al. 2009; Palumbi et al. 2007; Chataigner 2011).

The worked osseous material from the final Chalcolithic site of Godedzor is generally characterized by simple or even unplanned tools on the one hand and a few well made, multi-staged tool types typical of contemporary prehistoric sites in the region and even beyond into Anatolia (Choyke 2001; Choyke 2011: 86 – 98). Most of the 211 objects comprise tools made from the ribs and long bones of domestic ruminants such as cattle (*Bos taurus* sp.), caprinae and pig (*Sus* sp.?; tusk). Some sort of small-sized canid (*Canis dom.* or *Vulpes* sp.) and hare (*Lepus* sp.?) metapodia were carefully selected to produce a type of elongated bead. Wild animal bones, antler and teeth were used for manufacturing specialized objects used over long periods and which probably had symbolic content beyond their straightforward functional aspects. Hereafter "bone" will be used to refer to any hard-osseous material derived from animals including bone, antler and tusk.

The bone tools from Godezor were analysed using a handheld loop of 10 × and 20 ×. It would have been preferable to examine these objects under higher magnifications as well and had a longer time to study these objects. In the absence of better technical circumstances in the field, however, some conclusions or at least viable hypotheses about function, and typological analogy can still be drawn.

TRADITIONS IN BONE TOOL AND ORNAMENT MANUFACTURING IN THE PERIOD

Following the Neolithic, there is a general although not exclusive decline in the average quality, although not the importance, of bone tool manufacturing in daily life in the Near East and Europe (Sofaer et al. 2013: 482). Less effort seems to have been placed into bone tool production on the household level although the crafts these tools are used in (pottery, textile and hide working) remain of critical importance (Choyke 2000; 2005; Choyke and Bartosiewicz 2009). However, there are always exceptions to the rule. Preliminary work at the site of Ovçular Tepesi in northern Nakhchivan near the Armenian border presents a still unusual (at this stage of research) picture of a Late Chalcolithic group of possibly semi-nomadic pastoralists who seem to leave behind only complete, well made and planned caprinae long bone points that all could still have been used (Choyke and Christidou in preparation).

Perhaps the shift away from household production as the economic heart of life in villages and towns to production linked to complex trade and exchange mechanisms in societies of ever greater complexity and specialization may go some way to provide explanations for the more usual trend away from the carefully manufactured, planned tools of the Neolithic. However, there is a parallel tendency to produce or at least use ornaments and personal goods of an outstandingly high quality that were used intensively over long, possibly multi-generational time periods and which may well be produced along the serial production lines characteristic of market economies (*cf.* Choyke 2012a and 2012b). The use of metal tools (axes in particular) in the production of hard osseous materials only appears with antler production sequences in the later Chalcolithic although there is only a single instance of such an object found at the Godedzor site. A metal axe was definitively used to produce the general shape of a heavy-duty axe/hammer made from a red deer (*Cervus elaphus*) antler rack burr and beam. No clear evidence of metal use on other finished objects or the pieces of waste antler and bone (N = 2) came to light during excavations at Godedzor. Otherwise, the technical sequence (*châine d'operatoire*) remains largely unchanged from that employed in the wider region in earlier periods. The most common tools in the Godedzor assemblage are pointed tools or awls, mostly lacking their epipheseal ends, and rib spatulas. Both types were most probably multi-functional. A few special types point to the occupants of Godedzor being part of much wider technological horizons and traditions ranging from Anatolia and northern Iraq and Iran all the way to the Levant.

The bulk of the tools made at the settlement were made from animal bone available from the herd animals such as caprines and cattle. There are also objects made from the bones of local wild animals such as hare, wild canids, red deer and aurochs (Balasescu 2008; Balaescu 2009: 88 - 90). The shed antler for heavy-duty tools could have been picked up by the pastoralist herders on their way to this spring to early autumn occupation.

TYPOLOGY

Types were defined based on the shape of the working edge (also known as active end in the literature), size, raw material, form of the basis and gross manufacturing and macro-use wear traces observed at 10 × to 20 ×. The majority of these tools were probably designed to perform multiple tasks. Where appropriate, the bone tool typology adopted was originally used by Schibler (1980) for the Neolithic (Cortaillod Culture) lake-dwelling site of Twann on the Lake of Biel in western Switzerland. This typology has proved to be quite flexible because ultimately the types are based on the specific qualities of individual assemblages connected to species and skeletal elements and not simply the gross morphology of the objects themselves. There is only the need to add types which are unique to the assemblage. The disadvantage of purely form-based typologies is that names often include function. It has been shown conclusively that form does not necessarily equal function. Different forms can have identical functions and identical forms can have quite different uses or even multiple uses. Thus, analysis must revolve around manufacturing traditions and the social implications of the maintenance or loosening of such traditions rather than how these objects were used at a settlement and period (Stone 2011: 19). The main species/skeletal element-morphological types will be presented with little or no attempt to ascribe function beyond impressions of surface wears observed at relatively low, 20 ×, magnifications.

Pointed tools

Although the Chalcolithic in Europe is the period when the quality of workmanship in producing bone objects deteriorates, it is still striking how many Class II or Class I - II (selected raw material but low level of manufacturing) pointed bone tools were found at Godedzor. There is an especially large proportion of Class II pointed tools (Plate II. Figure 1) which are barely modified and apparently used for one-time tasks or only for short lengths of time. Although the "know-how" for producing grooved and split metapodial awls (N = 7) from caprinae (chiefly domestic sheep or goat in this site) still existed (Plate II. Figure 2) , the manufacturing choice tended to fall on longitudinally fractured caprine long bones that required little further modification to turn into usable if not very elegant tools.

Bevel-ended tools

The number of bevel-ended tools at Godedzor, as at the few other late Chalcolithic assemblages in the region where the worked bone material has been published, is low (N = 4). The bevel-ended specimens from Godedzor do not have well-preserved surfaces so together with the low numbers it is not possible to say anything about how they might have been used except to say that probably bone was the alternative raw material for tool types commonly made from other raw materials such as wood or even metal. There is one carelessly made (N = 1) scraper made from a split male wild boar lower tusk.

Object types possibly connected to textile production

The descriptions and measurements found here are based on individual tools used represent all the objects assigned to a particular type. It seems that textile production by these pastoralists, including spinning and weaving based on the bone and antler tools in this assemblage, took place during the late spring to early fall occupation at this upland settlement. Patterns of cheese and textile production has been recorded for other ethnographically attested transhumant pastoralist groups although these activities are by no means universal. The objects here have been ascribed function based on form and macro use wear patterns.

Rib spatula (Schibler type 12) (Plate III. Figure 1) These spatulas are made from the corpus of domestic cattle rib and represent a widespread type in time and space. The rib was split and the periostium on the outer compacta was scraped off with a flint tool. The spongiosa was abraded flat probably by using the abundant local stone slabs. The long edges of these spatulas were also rounded and refined by abrasion. Finally, the butt ends were abraded flat. These tools were usually intensively used and tend to be covered by use polish with rounding of the manufacture striations. Some are quite massive and robust. They may have suspension holes or notching at the broader end and appear to be used for extended periods of time, with evidence of repeated curation. These objects could have been used as a form of pin-beater in weaving on simple looms. It does not seem likely that all were made during this seasonal occupation. It can be expected that analogies to this type will be found at other contemporary sites in the region, perhaps in the territory of north-western Iran.

Fine needle (no Schibler type): N = 1 ; (Plate III. Figure 2) This single, delicate needle was made on the plantar surface of a caprine-size animal metatarsal. There is polish on the extreme tip. The needle possibly may have been used for sewing loosely woven textiles or leathers. Another eyed-needle may possibly have been used for nahl-binding (no Schibler type), (Plate Ⅲ. Figure 3) a type of single needle knitting normally used for making hats, gloves and socks, probably made from a caprine-sized tibia diaphysis. There are parallel examples of such tools from the contemporary Late Chalcolithic levels (4th millennium, level Ⅶ) at Arslantepe in Turkey. The nahl-binding needle is quite worn and may well have been brought with the pastoralist groups rather than being made at the settlement.

Spindle whorl (no Schibler type): N = 6; (Plate Ⅳ. Figure 1) These spinning tools were made from cattle, red deer, and aurochs (*Bos primigenius*) femur caput by cutting off the proximal epiphysis. The periostium was removed from the outer bone compacta, leaving flint scraping striations. The surface may have been polished. The hole was drilled though the center of the caput from both sides. This type is common in the final Chalcolithic levels of Arslantepe (level Ⅶ) as well as a possible occurrence in the Middle Bronze levels (Choyke 2000: 172, 178 – 179; Laureto, R. and Frangipane, M. 2010). These spindle whorls could be easily produced on the settlement or brought with these pastoralists on their seasonal round.

Weaving comb (no Schibler type): N = 1; (Plate Ⅳ. Figure 2) The comb is made from a red deer antler beam. A rectangular segment was extracted from the beam. No manufacture marks remain to show exactly how this was accomplished. A line was drawn to mark the extent of the teeth and then the teeth were cut out and a hole was drilled from both directions at the opposite end of object. A circle and dot design is incised below the hole and two triangles filled with incised oblique parallel lines lies above the line marking the length of the teeth. It is not possible to tell if the object was produced or decorated using flint or metal tools since antler may have been softened by soaking in water before working which may change the shape of the cuts. Experimentation on softened antler would be useful. The work is very precisely carried out. The long edges of the teeth are beveled and worn in a wavy curve. This is very similar to the use wear found on contact period rib weaving combs from Bolivia where continuity in use may be found (Choyke, 2014). Combs of a similar size, shape and macro use wear patterns were also found in the Middle Bronze Age sites of the Terre Mare in Northern Italy (Provenzano 2001a; 2001b). No parallels exist for this comb, but it is so well made and planned that its production tradition must be connected to Chalcolithic groups in the region whose material culture has not been published.

Other special tools

Heavy-duty hafted antler hammer/sleeve (Type Fa, Suter 1981): N = 1; Made from a piece of red deer antler beam below the crown of antler rack. The beam was sawn off below crown and 148 cm below, just above the place of the trez tine. The surface was chopped by a metal axe as shown by the acute "V"-shaped angle of the chop marks. The hafting hole was drilled from one side where the surface was prepared by chopping away the surrounding compacta. This specimen is relatively unused. A blade was probably inserted in one end of the sleeve, creating an axe type tool. Similar objects have been found at Arslantepe in both the final Chalcolithic and the Early Bronze age levels.

Pick or retoucher (no Suter or Schibler type): N = 1; A pointed tool made from the crown of a red deer antler rack. The rack was chopped around the base of crown which was then snapped off. The surface of the antler was not smoothed or modified. The rounded crown tine was battered at its tip. In general, there is relatively little antler tools at Godedzor beyond this object. The antler sleeve, the antler comb and some miscellaneous antler tine tips suggest that these tools were not manufactured on site, possibly because the raw material was not available and relatively heavy to transport. If these pastoralists headed up into the hills in late spring or early summer, then they would have had to carry the raw material with themselves since red deer stags drop their antlers in early spring. If the antler is not gathered relatively soon, it begins to degrade or is chewed on by the stags themselves and various rodents.

Cattle metatarsal with modified distal end (no Schibler type): N = 2; (Plate IV. Figure 3 – 5) Made on distal cattle or aurochs metatarsal. These strange tools look a bit like a modern monkey wrench. A broad rounded rectangular notch was cut in a dorsal-plantar direction through the center of the distal condyles. The inside of the rectangle is worn smooth in the better-preserved specimen. Their use is unknown. Another complete example and one half finished piece have been found at an Early Bronze Age site in Wadi Faynan in southern Jordan (Thomas Levy, personal communication). Since there is no consensus on what kind of work activity this object was made for, it is impossible to assess whether this activity was intense enough to have been produced this amount of wear on the inside of the distal condyle during a single season occupation at the settlement.

Pestle? (no Schibler type): N = 1; This tool is made from a slightly modified large domestic cattle tibia. The distal epiphysis is abraded flat and battered, suggesting its use as a crude pestle. It was probably made at the settlement.

Perforated first phalange (no Schibler type): (N = 8); (Plate V) This is a relatively common tool that were made from cattle, red deer or aurochs first phalanges. A crude hole is cut or sawn into the medullary cavity through the thin compacta on both sides. This is certainly a real type, but its use is obscure—especially given that the edge of the large central hole is usually broken. Only one specimen displays rounding on the edges of the hole as if a cord was passed though the hole. There are two unfinished specimens showing the tool was produced on-site during the seasonal occupation. Possibly these objects may have been used as a kind of simple, light weight. There are parallel examples from Late Chalcolithic levels (VII) at Arslantepe. Nevertheless, these objects were made for some activity at the settlement. One half made object, where the site of the perforation has been scraped out but not broken through into the medullary cavity, shows the local manufacture conclusively.

Thin tusk tool (Schibler 17): N = 1; (Plate VI. Figure 1) This finely made object is manufactured from the lower mandibular tusk of a male wild pig (*Sus scrofa ferrous*). The tusk was split, and the pulp cavity was scraped with obsidian or flint. The final form was created with abrasion. There was a bit of battering at the tip. Unusually for tusk tools, there was no defined long scraping edge so that this specimen could have been used in perforating special materials requiring a harder point. It may also have been a personal body ornament used in some kind of body piercing. Tusk is always a special material because wild boar is a dangerous game animal requiring considerable hunting prowess and bravery to kill—always assuming the tusk was not taken from a scavenged carcass. In any case, pig, whether wild or domestic, only appears in very low numbers throughout the archaeological sequence at Godezor (A. Balasescu in Kalantaryan et. al 2010: 47).

Long bone beads/closures (Schibler type 25): (N = 8); (Plate VI. Figure 2) These long bead-like objects were made from a mix of small canid, possible fox (*Vulpes* sp.?; Zsófia Kovács per. com.), and hare (*Lepus* sp.?; Zsófia Kovács per. com.) metapodia, one bird bone (*Avis* sp.?; Zsófia Kovács per. com.), and small mammal long bone diaphysis (possibly hare metapodial). The epiphyses were cut off and outer compacta deliberately polished producing fine perpendicular striations running around the circumference. The edges of the manufacturing striations were rounded from intensive use. These objects were either used as long, decorative beads or served as part of a closure for bags or clothing. It seems likely that they arrived on the site or were produced there for objects like clothing or bags intended for longer use than the seasonal occupation.

Cylindrical pendant/bead (s. f. # 14) (Schibler type 25): (N = 1); (Plate VII. Figure 1) The highly refined pendant/bead was made from a long bone diaphysis cut from near the long bone epiphysis of a cattle-sized animal. All manufacturing marks seem to have been removed by deliberate polishing into glossy surface. It was burned black. It seems likely to have been brought onto the settlement from some other location.

Perforated bear canine (Schibler type 23/2): (N = 1); (Plate VII. Figure 2) This pendant is made from a large adult brown bear (*Ursus arctus*) lower left canine tooth. The canine was deliberately split, so it would have lain flat on the skin or clothing. The hole was carved out (not drilled) through the root. This object was either a pendant or was sewn to clothing. It was broken once but continued to be used afterwards as the wear polish on the break testifies. The degree of polish and the deep honey brown color on the underside, particularly around the hole, suggest the tooth was in contact with leather or skin over a period of several decades. This estimation is based on the polish from skin contact on experimental beads worn continuously for five years (Duffy and Choyke in prep.).

Decorative pins (no Schibler type): (N = 6); (Plate VII. Figure 3) The decorative pins are made from large ungulate-size animal long bone. They are often curated. For some pins the quality of the workmanship and the perfectly round cross-section suggest they were made by specialists away from the pastoralist camp. On the other hand, the pins with heads from the Godedzor settlement were relatively heavily used since the manufacturing striations are rounded and polished. It seems most likely that such less well made, cruder pins, were probably manufactured at the settlement during the seasonal occupation. These objects are not very worn, which also suggests they were made to fill an immediate need and not brought onto the settlement on clothing.

Ring or clasp fitting (no Schibler type): (N = 1); (Plate VIII. Figure 1) The ring or clasp fixture may have been made from the long bone diaphysis of a wild caprine. A disk was cut out of the wall of a long bone. Part of the curve of the medullary cavity is retained in the slight S-profile of the ring. The object would not have been easy to wear as a finger-ring. A second disk-shape was cut out of the middle of the original disk, and then the tri-partite bezel was carved out of the upper part of the original disk. The shape is sophisticated and was probably deliberately polished. In any case, the ring/clasp was heavily worn from long-term use, so it is very unlikely that the ring was produced on-site. I know of no parallel finds anywhere else in the region, although there are contemporary parallels in the form of some bronze rings from Mediterranean Chalcolithic contexts.

Pigment container? (no Schibler type): (N = 1); (Plate VIII. Figure 2) GL = 18.3mm. The container was made from a left side femur diaphysis of possibly a bezoar goat (*Capra argrus*) or small adult cattle. It is too highly worked to identify to species. Both species are present in the faunal material from Godedzor. The two epiphyses were removed. The periosteum was scraped off the outer compacta leaving faint marks typical of a chipped stone tool. The entire surface is covered with incised lines alternating in oblique directions between bands of parallel lines perpendicular to the length. This piece is very similar to one produced at Tepe Gawra in northern Iraq, (level XI A, Late Chalcolithic 1, per. com. Chataigner). The incisions on the Godedzor container were also carved with a flint tool. Later in the Early Bronze Age, the lines on such objects are finer and most probably carved with metal tools. This kind of container is well known from the Early Bronze Age in the Near East, from the Aegean, the Levant to the Middle Euphrates region as well as western and eastern Anatolia. It becomes more common in archaeological materials from the Early Bronze Age II, reaching a peak in the Early Bronze Age IV of the Middle Eurphrates region (Hermann Genz 1996; 2003). However, four or five fragments of containers made from cattle femur diaphysis with incised line decoration come from the Final Chalcolithic and Early Bronze Age I at Arslantepe and the Greek Aegean, which suggests the type actually originated in more northerly parts of this vast region. Two femur containers were found *in situ* in rock-cut graves, one dated to EBA III and one to EBA IV, at Tell Shiyuk Tahtani in Syria. They were still filled with what was identified as blue-black kohl

(per. com Prof. Dr. Gioacchino Falsone).

Flute? (no Schibler type): (N = 1); This caprine femur diaphysis has two crudely drilled holes on the dorsal surface of the diaphysis one under the other. It may possibly be a flute or whistle of some sort. If indeed it was intended to be a flute or whistle, it was broken during manufacture.

DISCUSSION

The objects found during the excavations at Godozor were made from the skeletal elements of wild and domestic animal species found in the herds and wild animal populations found around the site. Several of the tool types found at Godedzor belong to broad tool manufacturing horizons that can be traced to the connections the Caucasus had with eastern Anatolia. These objects, such as the nahl-binding type needles, spindle whorls and the incised femur containers, are found in the Late Chalcolithic level VII of Arslantepe as well, levels with monumental architecture, but also dated to the mid-fourth millennium. The incised pigment container has exact parallels in the Late Chalcolithic levels at the tell site of Tepe Gawa in northern Iran. The cattle metatarsals with modified distal ends have parallels in the Levant, even further afield, although these specimens date to a slightly later time-period, to the Early Bronze Age of the region. These tools suggest that there were active cultural connections of some sort between eastern Anatolia and these small groups of transhumant pastoralists who moved seasonally into this upland area of southern Armenia, possibly from north-western Iran. The decorative pins were present in the Late Chalcolithic levels at Arslantepe but the forms are quite different. The pins from Godezor are crudely made in comparison and are morphologically dissimilar suggesting the models of such pins must lie in a different direction.

At the same time, there were clearly special objects such as the weaving comb, the nahl-binding needle, the nicely made spatula objects (possible weaving tools), the ornamented closing ring/clasp, the bone tubes made from hare or small canid metapodial diaphyses and perforated bear canine also appear to be brought onto the site from some other unidentified areas. It sugests these small groups of pastoralist also came into contact with settled peoples whose bone manufacturing traditions are not yet well known in the published literature. One might look for the source of these tools in the hard osseous material manufacturing traditions in larger urban contexts in the northern plains of Iran.

A final group of objects must also have been made at the site itself during the seasonal occupation. These are mostly *ad hoc* type, usually unplanned objects although the crudely made cattle first phalange with holes knocked through the diaphysis in a dorsal—ventral direction with rounding on the break represents a type which can also be found in Central-Eastern Anatolia but were certainly made at the settlement itself.

CONCLUSIONS

The worked osseous material from Godedzor can generally be divided into pointed tools which are not very carefully manufactured (Class I and Class I - II) and those probably produced on-site during the seasonal occupation and another group of tools and ornaments which are planned with multi-stage manufacture. With two exceptions, it appears Neolithic technologies employing chipped stone tools and abrasive stone materials were still employed to produce most of these objects. Some planned tools (Class I) are quite heavily worn and would have required skill and time to produce. It seems very likely that the amount of polish visible on these objects could not be produced in the four or five months spent by these pastoralists in their upland camp. The logical conclusion is that some of these ornaments and tools were brought with the pastoralists seasonally inhabiting this upland settlement, obtained from more settled communities, possible with specialized workshops, located in nearby villages. It would be worthwhile examining bone tool assemblages from contemporary settlements in both northwestern Iran and South Armenia to establish possible exchange networks for more elaborately manufactured objects. Such mutual dependencies based on regular trade and exchange of foodstuffs and other goods between nomadic pastoralists and local settled communities have been described for the region at large in ethnographies (Barth 1962, 1966). It is also natural that pastoralists on the move between plain and upland areas connected with and helped transfer new ideas, animals and even people in the Late Chalcolithic. Influences on their manufacturing tradition are clearly both local and unique and, at the same time, connected to large-scale manufacturing tradition horizons, sometimes even for very simple types such as the perforated cattle phalanges.

The bone industry at Godezor appears to have been geared toward craft activities such as fiber production (spindle whorls) and perhaps weaving as well as indicated by the spatulas, the weaving comb and even the perforated phalanges which may have been used as weights. Many of the decorative objects appear to have come on other objects such as clothing or bags as parts of the closures or as ornaments. The heavy-duty antler tools and antler picks may have been used to move earth or cut small trees or bushes. Similarly, the pointed bone tools (awls) may have been used in a variety of materials, for example, making hide objects or in sewing. Without high magnification examination and proper experimentation all the functional attributions based on macro-wear and gross morphological characteristics must be treated with caution. The more elaborated objects generally display wear indicating their use over a longer time while the *ad hoc* or locally made objects generally exhibit use wear suggesting they were used for longer periods than the seasonal occupation of the settlement.

The Late Chalcolithic bone tool manufacturing traditions and their use and meaning in these small-scale scale remains difficult to assess based on a single isolated site material. Hopefully, further research will shed light on the material culture of these more ephemeral populations and their connections with larger settled groups. With more data, scholars can begin to assess mutual socioeconomic dependencies in this period in this part of the world.

REFERENCE

Avetisyan, P., Kalantaryan, I. Palumbi, G., Chataigner, C., Balasescu, A. and Horespyan, R. (2009) *L'établissement Chalcolithique de Godezor et les relations entre Petit Caucase et nord de l'Iran*. MAEE (Ministère des Affaires Étrangères et Européennes).

Chataigner, C. (2016) Environments and societies in the southern Caucaus during the Holocene. *Quaternary International* 395, 1 – 4.

Chataigner, C., Avetisyan, P., Palumbi, G., Uerpmann, H-P. (2010) Godezor, a Late Ubaid-related settlement in the southern Caucaus. In: R. Carter and G. Phillips (eds.) *Beyond the Ubaid—Transformation and Integration in the Late Prehistoric Societies of the Middle East.* SAOC 63, 377 – 394. Chicago, the Oriental Institute of the University of Chicago.

Choyke, A. M. (1997) The bone manufacturing continuum. *Anthropozoologica* 25 – 26, 65 – 72.

Choyke, A. M. (2000) Bronze Age antler and bone manufacturing at Arslantepe (Anatolia). In: M. Mashkour, A. M. Choyke, H. Buitenhuis (eds.) *Archaeozoology of the Near East* IVA. ARC Publication 32, 170 – 183. Groningen, the Netherlands.

Choyke, A. M. (2001) A quantitative approach to the concept of quality in prehistoric bone manufacturing. In: H. Buitenhuis, W. Prummel (eds.) *Animals and Man in the Past.* ARC Publication 41, 59 – 66. Groningen, the Netherlands.

Choyke, A. M. (2005) Bronze Age bone and antler working at the Jászdózsa-Kápolnahalom tell. In: H. Luik, A. Choyke, C. Batey and L. Lõugas (eds.) *From Hooves to Horns, from Mollusc to Mammoth: Manufacture and Use of Bone Artifacts from Prehistoric Times to the Present.* Proceedings of the 4th meeting of the (ICAZ) Worked Bone Research Group, Tallinn, August 2003, 129 – 156. Tallinn: Ajaloo Instituut.

Choyke, A. M. (2011) Étude des artefacts en os. In: C. Chataigner (ed.) *Report Scientifique.* MAEE (Ministère des Affaires Étrangères et Européennes), 86 – 98.

Choyke, A. M. (2012a) The bone workshop in the church of San Lorenzo in Lucina. In: O. Brandt (ed.). *San Lorenzo in Lucina. The transformations of a Roman quarter.* Series: Acta Instituti Romani Regni Sueciae, 4, vol 61, 335 – 346. Stockholm.

Choyke, A. M. (2012b) Skeletal elements from animals as raw materials. In: *Bone Objects in Aquincum.* Az Aquincum Múzeum Gyűjteménye 2. (The Collections of the Aquincum Museum 2), 43 – 53. Budapest.

Choyke, A. M. (2014) Chapter VII. Worked Skeletal Materials from Paria. In: J. Gyarmati and C. Condarco Castellón (eds.) *Paria La Viexa, Pre-Hispanic Settlement Patterns in the Paria Basin, Bolivia, and its Inka Provincial Center*, 113 – 118. Museum of Ethnography: Budapest.

Choyke, A. M., and Bartosiewicz, L. (2009) Telltale tools from a tell: Bone and antler manufacturing at Bronze Age Jászdózsa-Kápolnahalom, Hungary. *Tiscium* XX, 357 – 376.

Genz, H. (1996) Ritverzierte Knochenhülsen vom Tell el-Abd. *Dmaszener Mitteilungen* 8, 81 – 82.

Genz, H. (2003) *Ritzverzierte Knochenhülsen des dritten Jahrtausends im Ostmittelmeerraum. Ein Beitrag zu den frühen Kulturverbindungen zwischen Levante und Ägäis*. Abhandlungen des Deutschen Palästina-Vereins 31. Wiesbaden, Harrassowitz Verlag.

Laureto R. and Frangipane M. (2010) Textile tools and textile production. The archaeological evidence of weaving at Arslantepe. In: M. Frangipane (ed.) *Economic Centralisation in Formative States. The Archaeological Reconstruction of the Economic System in 4th Millennium Arslantepe*, SPO 3, 275 – 285. Roma, Sapienza Università di Roma.

Kalantaryan, I., Avetisyan P., Palumbi, G., Chataigner, Le Mort, F., Poumare'h, M. and Balasescu, A. (2010) *L'établissement Chalcolithique de Godezor*. MAEE (Ministère des Affaires Étrangères et Européennes).

Kalantaryan, I., Avetisyan P., Palumbi, G., Chataigner, C., Gasparyan, B., Balasescu, A. and Hovsepyan, L. (2008) *L'établissement Chalcolithique de Godezor*. MAEE (Ministère des Affaires Étrangères et Européennes).

Palumbi, G. and Chataigner, C. (2014) The Kura-Araxes culture from the Caucaus to Iran, Anatolia and the Levant: Between unity and diversity. A synthesis. *Paléorient* 40(2), 247 – 260.

Palumbi, G., Hovsepyan, R. and Chataigner, C. (2007) *L'établissement Chalcolithique de Godezor dan la sud-est de Petit Caucase*. MAEE (Ministère des Affaires Étrangères et Européennes).

Provenzano, N. (2001a) Artisanat des matières dures animales en milieu terramaricole: état de la question, In: S. Méry, C. Karlin and A. Averbouh (eds.) Systèmes de production et de circulation, Cahier des Thèmes transversaux ArScAn, MAE-Nanterre—Thème transversal n 3, Les moyens d'action sur la matière (production, transformation, échanges des produits), 152 – 156.

Provenzano, N. (2001b) Worked bone assemblages in Northern Italy Terremares: A technological approach. In: A. M. Choyke and L. Bartosiewicz (eds.) *Crafting Bone: Skeletal Technologies through Time and Space*. Proceedings of the 2nd meeting of the (ICAZ) Worked Bone Research, Budapest, September 1999. British Archaeological Reports International Series 937, 93 – 103. Oxford, Archaeopress.

Schibler, J. (1981) *Typologische Untersuchungen der cortaillodzeitlichen Knochenartefakte. Die neolithischen Ufersiedlungen von Twann*, Band 17. Schriftenreihe der Erziehungsdirektion des Kantons Bern, herausgegeben vom Archäologischen Dienst des Kantons Bern, Staatlicher Lehrmittelverlag (Bern).

Sofaer, J., Jørgensen, L. B. and Choyke, A. (2013) Craft production: Ceramics, textiles and bone. In: A. Harding and H. Fokkens (eds.) *Oxford Handbook of the Bronze Age*, 469 – 492. Oxford, Oxford University Press.

Suter, P. J. (1981) *Die Hinchgeweihartefakte der Cortaillod-Schichten*. Bern: Staatlicher Lehrmittelverlag. Die neolithischen Ufersiedlungen von Twann, Bd 15.

Bird Bone Double Pipe from the 7th Century Avar Cemetery of Szegvár-Szőlőkalja (South-East Hungary)

Erika Gál

Institute of Archaeology, Research Centre for the Humanities,
Hungarian Academy of Sciences, Hungary

Abstract: This paper presents the zooarchaeological analysis of a double pipe found in a female grave in the Early Avar period (7th century AD) cemetery of Szegvár-Szőlőkalja in South-East Hungary. This type of musical instrument is a characteristic grave donation (usually to burials of men) found in Avar period cemeteries in the eastern part of Hungary. The specimen under study was made from the matching pair of tarsometatarsi of a crane. This graceful species with its particular sound once was a common breeding bird on the Great Hungarian Plain. Its long and straight leg bones seem to have been an ideal raw material for manufacturing double pipes.
Keywords: Avar period, cemetery, double pipe, crane bone, Hungary

INTRODUCTION

The Avar Khanate was established in the Carpathian Basin in 567 by the Avars, a group of people of uncertain Central Asiatic origins. At the wake of the Migration Period, this largely pastoral society was the first to integrate the heterogeneous population of the Carpathian Basin into a single empire for some 250 years, marking the transition from Late Antiquity to the Early Middle Ages. The multiple origins of Avars are reflected in their diverse burial customs including funerary pyre, grave pits with a side niche, and simple inhumation. Members of what could be called aristocracy and military leaders were often buried with their horses and weapons. Personal items of the buried people such as sabretaches, combs, earrings, crosses, weights and cups as well as drink and food donation for the journey to the afterworlds were placed into the graves (Vida 2003: 304 – 305).

The double pipe (inventory number 84.1.174) found in the 7th century Avar cemetery at Szegvár-Szőlőkalja was revised from an osteological point of view at the Móra Ferenc Museum, Szeged, in 2005 at the request of late Lívia Bende. That time I was able to study a similar musical instrument (inventory number 95.8.1) excavated from the Avar Period cemetery at

Felgyő, Ürmös-tanya as well (Balogh 2010: 337, Figure 63.1).

The flute from Szegvár-Szőlőkalja (Figure 1) was uncovered during field work led by Katalin Hegedűs in 1979. A total of 93 graves dated to the Early Avar period were found during the excavation, whose majority were oriented NW-SE. According to wood residue found at the bottom of some graves and the positions of human skeletons, over 80% of the graves have been indentified as coffin burials. Although the cemetery seems to have contained burials of common people, it was rich in grave goods, especially regarding pottery (67 vessels) and food donations represented by animal bones. The latter included remains of cattle, sheep and goat, and poultry. In addition, a number of eggs of possible cultic significance were also found (Hegedűs 1979).

The completely parallel double pipe with five sound holes on each tube was placed in grave no. 109, which included the remains of an adult woman. The anthropological study of this skeleton revealed a slight form of metabolic disease on the skull (Farkas et al. 1993: 13). The grave was oriented Northwest-Southeast. In addition to the musical instrument, other offerings such as accessories (a small and a large iron buckle, and an iron knife), pottery and food donation represented by a cattle vertebra were also found in the grave (Bende 2005: 169, Figure 139).

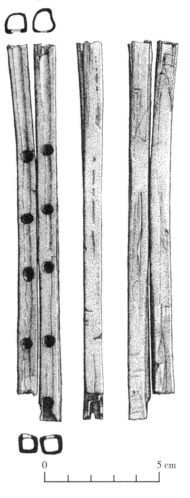

Figure 1. Dorsal, side and plantar views (left to right) of the double pipe from Szegvár-Szőlőkalja (after Bende 2005, Figure 139).

Besides the field report already cited, other papers have also dealt with this interesting find. It has been stated that the flute was made from bird bones (Kürti and Menghin 1985: 70 – 71, Figure 70) even the skeletal part used and the bird species have been correctly appraised by the music player and researcher József Kozák. He also noted that the type of musical instrument and the distance between its second and third holes indicates major third pentachord scale (Kozák 1997: 196 – 197, Figure 2). The music teacher and ethnologist György Csajághy stated that the flute could not be played due to its poor preservation (Csajághy 2003). However, the first ever comprehensive archaeozoological study of the find, including the use of a comparative skeleton, is presented in this study.

RESULTS

This analysis confirmed that the double pipe was indeed made from the shafts of both tarsometatarsi from crane (*Grus grus* Linnaeus, 1758), just like the pipe excavated from Felgyő.

As it looks evident for the first sight, the pipes were made from a matching pair of bones from the same individual. The instrument was probably blown from the proximal ends of the two bones. Thorough osteological investigations showed that the longer tube of the flute, which is the left side piece from the viewpoint of the player, was made from the right leg of the crane, while the shorter tube was produced from the left leg of the same bird (Figure 2). This inverted position means that the two bones were attached to each other with their "outer", lateral surfaces. The bone shafts were cut off at the metaphyses, right below the more distally located *foramina vascularia proximalia* in both cases (Baumel 1979: 121).

The musical instrument is generally poorly preserved. The plantar as well as the medio-lateral surfaces are especially broken. Owing to the restoration, the upper (proximal) ends slightly depart from each other. The lower (distal) ends of both tubes are broken. The pipe has been complemented at several points during restoration.

The rectangular cross section of tubes is provided by the anatomical shape of tarsometatarsus in crane. Originally, the bone has longitudinal edges on the dorsal and plantar surfaces alike, as well as on the medial and lateral sides. It is evident that these lines have been carved away on all four sides. This operation was carried out in a longitudinal direction but not always parallel to the bone shaft. According to the marks left on the tubes this stage of work was completed by a number of short and sometimes irregular movements. Each grinding mark shows some very thin and quite deep parallel lines. Only a very few carving marks run perpendicular to the shaft.

Both tubes hold four sound holes, the fifth pair is broken on each pipe. The proximal end of the left tube is incomplete. The corresponding end of the right pipe is relatively well preserved. It shows, however, two-three shallow cut marks, which indicate that at first the size of the tube was wrongly considered. The natural curvature of tubes allows us to establish that the sound holes were drilled on the dorsal surfaces of bones. Previous markings indicating the location of holes have not been noted. These may have disappeared during manufacturing. The plantar surface of bones is intact but poorly preserved showing numerous cracks.

The complete length of the right-side tube is 147.1 mm. The width of the proximal end is 9.4 mm while its depth is 8.6 mm. The distal ends are difficult to measure because the tubes are tightly fixed to each other. Nevertheless, they do not considerably differ from the sizes of the proximal ends. Owing to the modern reparations, the thickness of bone walls is also complicated to estimate, but it varies between 1.0 – 1.5 mm.

The left side tube (158.2 mm) is somewhat longer than the right-side pipe. At least the half of the perimeter of the broken 5th sound hole on its distal end is preserved. The width of the proximal end is 10.3 mm and its depth is 9.6 mm. Corresponding measurements on the distal ends measure 9.3 mm and 9.4 mm, respectively. Both tubes have been described and their general cross-sections illustrated by György Csajághy in his book written on the parallel double pipe from Felgyő, Ürmös-tanya (Csajághy 1998: 29).

The margins of the sound holes are intact and show regular circles in most cases. Their repeated measurement at every revision could damage the holes. In spite of this possibility, we tried to give exact measurements for the diameters of holes, which vary between 3.9 – 4.9 mm. The distances between the sound holes measured from their central point range from 23.5 to 30.7 mm. The drilled surfaces of the tubes are flat.

The clearly visible root etching on the drilled surface of the pipe would suggest that it was placed in the grave in a functional position, i.e. with the sound holes facing "upwards". The same pattern of root marks was noted on the pipe from Felgyő, Ürmös-tanya, although the archaeological report indicates just the opposite, namely that this latter instrument was placed with the holes facing down (Csajághy 1998: 28 – 29). This contradiction raises the question whether the grave had been disturbed a short time before the excavation was made by specialists.

DISCUSSION

So far 12 Avar Period double pipes have been reviewed (Keszi 2016: 98), described and some even experimentally played by the musicologists György Csajághy and József Kozák (Kozák 1997; Csajághy 1998, 2003). They originate from the next sites: Szegvár-Szőlőkalja, Felgyő-Ürmös-tanya, Kunmadaras-Benzinkút-Újvárosi temető, Tiszagyenda-Búszerző, Jánoshida-Tótkérpuszta, Alattyán-Tulát, Rácalmás-Rózsa-major, Bonyhád-Hidasi téglagyár, Tatabánya-Alsógalla, Balatonfűzfő-Szalmássy-telep, and Keszthely-Dobogó. They were all found in burials with the exception of the pipes from Kunmadaras-Benzinkút-Újvárosi temető and Tiszagyenda-Búszerző, which came to light from dwellings (Molnár 2010; Keszi 2016: 98). The best-known double pipes are those from Jánoshida-Tótkérpuszta, Alattyán-Tulát, and Felgyő, Ürmös-tanya, presented in longer publications (Bartha 1934; Kovrig 1963; Csajághy 1998). The closest similarity to the musical instrument from Szegvár-Szőlőkalja is shown by the two flutes from Alattyán-Tulát (graves no. 285 and 477) and the flute from Tiszagyenda-Búszerző, respectively. All these double pipes were made from crane tarsometatarsi and are completely parallel pipes with five sound holes (Kovrig 1963: 173, Plate 76; Molnár 2010).

The pipe found at Felgyő, Ürmös-tanya is also ranked among these wind instruments, but it has six holes instead of five. Nevertheless, it has been stated that the fifth pair of holes was pierced by mistake and most probably was filled during the time the pipe was played (Kozák 1997: 196 - 197, Figure 2). Most of the other Avar Period, completely or semi-parallel double pipes also were produced from skeletal parts of crane, usually from tibiotarsi and tarsometatarsi (Kozák 1997: 196; Keszi 2006). However, the semi-parallel double pipe from Jánoshida-Tótkérpuszta was made from ulnae (Bartha 1934: 18, Plate 1. Figure 4; Kozák 2002: 26, Figure 10). The characteristic straight shape and rectangular cross-section of the bone shaft in crane tibio-and tarsometatarsi make these bones especially suitable for fitting and fastening the two tubes to each other. A simple flute excavated from the Roman Period site Großsachsenheim (Germany) also was made from a skeletal part of crane (Schallmayer 1994: 73, Figure 3).

Crane is one of the biggest and most beautiful birds of grassy lowlands and meadows. Its elegant movements and feathers as well as the peculiar dance performed during the mating season have made people interested in this bird for a very long time. Crane has permanently bred in Hungary until the 19th century. Since then it is mostly an autumn-spring passage species (Herman 1901; Hume 2003: 160).

Crane has been a much-liked prey animal all over the Carpathian Basin since the Neolithic (Gál 2004: 280 - 286, Tables 1 - 4). Many crane finds were excavated, among others, from Bronze Age sites (Gál 2013). During the Middle Ages, it was a fashionable to keep tamed cranes in the courts of high society just for pleasure or the use of feathers in decorating dresses and other objects. The Hungarian scientist Ottó Herman writes—with a touch of exaggeration—that: "…There was not a manorial courtyard in Hungary where a tamed crane earnestly and with the dignity of an heraldic bird did not live among the poultry, and danced sometimes by alternating its feet, raising its head and flapping its wings…" (Herman 1901).

An interesting summary on the exploitation of cranes as food resource and symbol, based on archaeological and ethnographic data, has recently been published (Bartosiewicz 2005). Crane bone manufacturing, however, fell beyond the focus of that paper. Bird bones have been well-tried basic materials for making different wind instruments even since the Palaeolithic period (Soler-Masferrer-Garcia-Petit 1995; Omerzel-Terlep 1997). Several long bones such as the ulna, tibiotarsus and tarsometatarsus of large birds including crane are first-class raw materials for producing pipes and flutes due to the typical shape as well as the thin-walled structure of these pneumatic skeletal parts. The oldest playable flutes, excavated from the Early Neolithic (7000 - 5700 BC) site Jiahu in China were made from red-crown crane (*Grus japonensis* Müller, 1766) ulnae. The six complete specimens have five, six, seven or eight holes (Zhang et al. 1999). The review of recently excavated and described bird bone artefacts from Hungary also includes musical instruments (Gál 2005).

Nevertheless, the Avar Period double pipes are rare finds, which do not resemble any bird bone instruments from other periods. The relative frequency as well as similarity of the completely and partially parallel double pipes indicates the importance of these wind instruments as well as the possibly special meaning of melodies played to the people living at the end of 7th century.

ACKNOWLEDGEMENTS

The late archaeologist, Lívia Bende, invited me to study the wind instruments and provided information regarding their earlier descriptions. Gábor Lőrinczy kindly provided access to the relevant part of Bende's unpublished PhD dissertation. László Bartosiewicz is thanked for preparing the illustrations and correcting the English text of this paper.

REFERENCES

Balogh, Cs. (2010) A Felgyő, Ürmös-tanyai avar kori temető(The Avar Cemetery at Felgyő, Ürmös-tanya). In: Cs. Balogh and Klára P. Fischl (eds.) *Felgyő, Ürmös-tanya*. Monumenta Archeologica 1, 185 – 381. Szeged, Móra Ferenc Múzeum.

Bartha, D. (1934) *Die Avarische Doppelschalmei von Jánoshida*. Archaeologica Hungarica 14. Budapest, Magyar Történeti Múzeum.

Bartosiewicz, L. (2005) Crane: Food, pet and symbol. In: G. Grupe and J. Peters (eds.) *Feathers, grit and symbolism. Birds and humans in the ancient Old and New Worlds*. Documenta Archaeobiologiae 3, 259 – 269. Rahden/Westf., Verlag Marie Leidorf GmbH.

Baumel, J. J. (1979) Osteologia. In: J. J. Baumel (ed.) *Nomina Anatomica Avium*, 53 – 121. London, New York, Toronto, Sidney, San Francisco, Academic Press.

Bende, L. (2005) *Temetkezési szokások a Körös-Tisza-Maros közén az avar kor második felében* [Burial customs in the Körös-Tisza-Maros region during the second part of the Avar Period]. PhD dissertation. Budapest, Eötvös Loránd University, Faculty of Humanities.

Csajághy, Gy. (1998) *A felgyői avar síp és történeti háttere* [The Avar pipe of Felgyő and its historical background]. Budapest, Püski.

Csajághy, Gy. (2003) Még egyszer az avar sípokról [One more time about the Avar flutes]. *Eleink* 2(2), 43 – 71.

Farkas, L. Gy, Marcsik, A. and Oláh, S. (1993) Történeti idők embere Szegváron [The man of historical times in Szegvár]. *Anthropológiai Közlemények* 35, 7 – 37.

Gál, E. (2004) The Neolithic avifauna of Hungary within the context of the Carpathian Basin. *Antaeus* 27, 273 – 284.

Gál, E. (2005) New data to the bird bone artefacts from Hungary and Romania. In: H. Luik, A. M. Choyke, C. E. Batey and L. Lõugas (eds.) *From Hooves to Horns, from Mollusc to Mammoth. Manufacture and Use of Bone Artefacts from Prehistoric Times to the Present*. Muinasaja teadus 15, 325 – 338. Tallin, Archaeological Depertment of the Institute of History-Chair of Archaeology of the University of Tartu.

Gál, E. (2013) Bird bone remains from Bronze Age settlements in the Carpathian Basin. In: M. Vicze, I.

Poroszlai and P. Sümegi (eds.) *Hoard, Phase, Period. Round Table Conference on the Koszider Problem*, 193 – 204. Százhalombatta, Matrica Múzeum.

Hegedűs, K. (1979) A Szegvár Szőlőkaljai leletmentő ásatás [The rescue excavation at Szegvár Szőlőkalja]. *Múzeumi Kutatások Csongrád megyében* 1979, 59 – 66.

Herman, O. (1901) *A madarak hasznáról és káráról* [*On Birds' Benefit and Harm*]. Arcanum PC CD-ROM.

Hume, R. (2003) *Madárvilág Európában* [*Birds of Europe*]. Budapest, Panemex-Grafo.

Keszi, T. (2006) Avar kettős síp Rácalmásról [Avar Period double pipe from Rácalmás]. *Az Intercisa Múzeum Évkönyve* 1, 97 – 99.

Kovrig, I. (1963) *Das awarenzeitliche Gräberfeld von Alattyán*. Archaeologica Hungarica 40. Budapest, Akadémiai Kiadó.

Kozák, J. (1997) Kettétört csontsípszár a Bijelo brdoi avarkori temetőben [Broken bone pipe from the Avar Period cemetery of Bielo Brdo]. *Communicationes Archaeologicae Hungariae* 1997, 195 – 203.

Kozák, J. (2002) Sípok és népek az avarkori Magyarországon [Flutes and people in Avar Period Hungary]. *Turán* 5(6), 19 – 30.

Kürti, B. and Menghin, W. (1985) Katalog. In: W. Meier-Arendt (ed.) *Awaren in Europa. Schätze eines asiatischen Reitervolkes* 6. – 8. *Jh.*, 24 – 86. Frankfurt am Main, Museum für Vor-und Frühgeschicthe.

Molnár, E. (2010) A tiszagyendai avar csontsíp [An Avar bone flute from Tiszagyenda]. In: J. Gömöri and A. Kőrössi (eds.) *Csont és bőr: Az állati eredetű nyersanyagok feldolgozásának története, régészete és néprajza* [*Bone and Leather. History, Archaeology and Ethnology of Crafts Utilizing Raw Materials from Animals*], 93 – 98. Budapest, Magyar Tudományos Akadémia-VEAB Soproni Tudós Társasága.

Omerzel-Terlep, M. (1997) A typology of bone whistles, pipes and flutes and presumed Palaeolithic wind instruments in Slovenia. In: I. Turk (ed.) *Mousterian "Bone Flute" and Other Finds from Divje Babe I Cave Site in Slovenia*. Opera Instituti Archaeologici Sloveniae 2, 199 – 218. Ljubljana, Institut za Archaelogijo.

Schallmayer, E. (1994) Die Verarbeitung von Knochen in römischer Zeit. In: M. Kokabi, B. Schlenker and J. Wahl (eds.) *"Knochenarbeit". Artefakte aus tierischen Rohstoffen im Wandel der Zeit*. Archäologische Informationen aus Baden-Württemberg 27, 71 – 82. Stuttgart, Gesellschaft für Archäologie in Württemberg und Hohenzollern.

Soler Masferrer, N. and Garcia Petit, L. (1995) Un probable xiulet paleolític a Davant Pau (Serinyà, el pla de l'Estany). In: J. Padró Parcerisa (ed.) *Cultures i medi. De la Prehistòria a l' Edat Mitjana. 20 anys d'arqueologia pirinenca. Col · loqui Internacional d'Arqueologia de Puigcerdà*, 195 – 206. Puigcerdà, Institut d'Estudis Ceretans.

Vida, T. (2003) The Early and Middle Avar Period (568 – turn of the 7th – 8th centuries). In: Zs. Visy (ed.) *Hungarian archaeology at the turn of the millennium*, 302 – 307. Budapest, Ministry of National Cultural Heritage-Teleki László Foundation.

Zhang, J., Harbottle, G., Changsui, W. and Kong, Z. (1999) Oldest playable musical instruments found at Jiahu early Neolithic site in China. *Nature* 401, 366 – 368.

Middle Bronze to Early Iron Age Bone Tools from Eastern Germany

Hans Christian Küchelmann

Knochenarbeit, Speicherhof 4, 28217 Bremen, Germany

Abstract: A group of five bone artefacts from the Middle Bronze to Early Iron Age settlement site of Kemnitz near Berlin is presented. The artefacts are simple objects but were obtained from at least three animal species. Manufacturing traces and use wear are described and possible interpretations were tested with experiments. One object allowing further considerations is a bone spearhead, which is discussed in context with comparative finds.
Keywords: Bronze Age, Early Iron Age, Eastern Germany, bone artefacts, bone spearhead

Remains of a Middle Bronze to Early Iron Age settlement were discovered during a rescue excavation from June to August 2001 in Kemnitz, district Potsdam-Mittelmark, Brandenburg about 50 km East of Berlin. The excavation covered an area of roughly 120 m^2 and did not reach the final edge of the site. Within this relatively small area, 15 stone fireplaces, two campfires constructed from pottery shards and 45 pits were discovered (Figure 1). Charcoal samples have been radio-carbon-dated to 800 BC although the pottery finds indicate an occupation lasting from 1100 to 600 BC (Buck 2002: 71).

About 700 animal bone fragments were found distributed between 44 features. The faunal remains are now stored in the archives of the Brandenburgisches Landesamt für Denkmalpflege und Archäologisches Museum (BLADAM). A small sample of these bones (53 specimens) was identified with the help of the reference collection of the Archäologisch-Zoologische Arbeitsgruppe (AZA), Zentrum für Baltische und Skandinavische Archäologie in Schleswig. The sample contained bones from domestic pig (*Sus domesticus*), sheep or goat (*Ovis / Capra*), domestic cattle (*Bos taurus*), hare (*Lepus europaeus*) and a small carnivore tooth-fragment, probably from a dog (*Canis familiaris*). Among the animal bone finds were five worked bones (Table 1) to be discussed in this paper (see also Küchelmann 2002).

Three tools are of a simple type combining a grip and a tip part. Tool no. 1 was cut from long bone compacta of a cattle to red deer sized animal (length 58 mm, width 6 mm). The raw material can not be identified morphologically to species level. The grip is roughly shaped with seven facets, four of which show traces of rough abrasion perpendicular to the tool-axis. The abrasion marks are comparable to the marks left by a modern wood-rasp (cut 2). The tip is more elaborately worked and polished all around. Unfortunately, the point was broken during excavation (Plate IX. Figure 1).

Figure 1. Excavation area of a Middle Bronze to Early Iron Age settlement at Kemnitz, district Potsdam – Mittelmark, Brandenburg, Germany (Photo: Nordholz).

Table 1. Bone artefacts from Kemnitz, Germany.

No.	Inventory no.	Species	Skeletal Element	Interpretation
1	BLADAM 2001 – 779/18/3/11	cattle to red deer sized mammal	?	pointed tool
2	BLADAM 2001 – 779/38/3/24	cattle to red deer sized mammal	?	pointed tool
3	BLADAM 2001 – 779/53/3/46	*Sus domesticus* · pig	fibula	pointed tool
4	BLADAM 2001 – 779/41/3/34	*Ovis / Capra* · sheep or goat	tibia	projectile point
5	BLADAM 2001 – 779/38/3/23	*Lepus europaeus* · hare	tibia	support tool

Worked bone no. 2 derives from a cattle to deer-sized animal long bone as well, but is worked much more precisely. This 51 mm long tool is evenly ground as if its surface had been worked with modern sandpaper with a grain size of 60. The grip is rectangular in cross-section (6 × 5 mm) with two convex facets and two planar facets. The latter facets fit perfectly between thumb and forefinger. The tip is triangular in cross-section and points to one of the convex facets. The point shows sign of polish, probably connected to use-wear (Plate IX. Figure 2). Cut from the distal half of a left pig-fibula, tool no. 3 needed minimum effort to produce a useful shape. The fibula simply had to be sharpened proximally and was ready to be used for work (length 42 mm, width 5 × 3 mm; Plate IX. Figure 3).

All these tools could be used for any kind of work where a hole or embossing on a soft material is desired. Similar tools were used from the Neolithic to the Middle Ages in different cultures (see for instance Becker 2001: 129 – 148; Beech 1995: 110, 140; Choyke 1996: 307, 309, Figure 3; Kokabi et al. 1994: 63 – 67; Maldre 2001: 21, 28; Schwarz-Mackensen 1976; Teegen 1999: 34, Figure 55, no. 3290, 4220).

A replica of tool no. 2 was made from a fragment of cattle metapodium using modern tools (saw, rasp, file, sandpaper with grain-sizes 80 and 100) to test different ways it could have been used. Air-dried, unburnt pottery was chosen as one application because pottery finds from the site include shards decorated with engraved lines (Buck 2002). The shape of the tip, being sharp but broadening to a diameter of 5 – 6 mm within a distance of 8 mm, on the other hand suggested the possibility of an application on a more robust material. Therefore, it was tested if the tip would be sharp and stout enough to perforate leather. The result was that the replica could be nicely employed in perforating leather (Figure 2. a) as well as for decorating pottery either with engraved lines or with triangular imprints (Figure 2. b). However, as already mentioned, various other functions are possible and the question, which kind of action was performed with the tool can only be answered by a series of experiments with identical tools employed repeatedly on different materials with a subsequent microscopic use-wear analysis of the artefact and the replicas.

Figure 2. Testing the suitability of bone tool no. 2 (Photos: Küchelmann).
a. leather perforation.   b. pottery decoration.

An artefact of a different type (no. 4) was made from a right tibia of sheep or goat. The proximal epiphysis was removed, the diaphysis is cut obliquely beginning at the lateral side, two thirds of the way down the shaft, forming an asymmetrical tip on the medial diaphysis directly above the base of the distal epiphysis. The cutting facet is 40 mm long, the remaining length of the object is 102 mm. Six facets are visible. All show traces of rough abrasion perpendicular to the axis of the bone. No effort was made to smooth the surface. The tip was broken in the past (Plate X. Figure 1).

Obliquely cut pointed bone shafts like this one are common artefacts in Northern Germany, Scandinavia, Poland and the Eastern Baltic. The main period of their use is the Pre-Roman Iron Age (5th – 1st century BC), a smaller amount has been found at Late Bronze Age (1st half of the 1st millennium BC) and Roman Iron Age sites (1st – 3rd century AD). The majority of the artefacts has been made from long bone shafts of ovicaprids, mainly tibiae (Bräunig 2002; Luik 2013: 26 – 27; Raddatz 1954; Schoknecht 1983). Most informative are the finds from the Iron Age (3rd century BC) bog sacrifice site of Hjortspring in Denmark. Altogether 169 spearheads were discovered at this site among other finds. Twenty-six of these spearheads were made from long bone diaphysis segments, most of which are tibiae of sheep or goat like the finds from Kemnitz. They were manufactured in the same manner and were of a comparable size (95 to 130 mm; Plate X. Figure 2). Some of the Hjortspring spearheads were mounted on approximately 2 m long ash wood shafts either with rivets or glue (Kaul 1988). Further 19 bone spearheads are known from another bog sacrifice site at Krogsbølle, Denmark (Schoknecht 1983: 48). Schoknecht (1983) also presents 55 specimen found in Mecklenburg-Vorpommern, Germany, 22 of which were provided with rivet holes. Bräunig (2002) published a catalogue with finds from all over Europe, including distribution maps. More recent finds from Germany, not mentioned by Bräunig (2002), are reported from Hitzacker (n = 1; Becker 2009: 87 – 88) and Bentumersiel (n = 2, one with a rivet hole; Küchelmann 2011: 62 – 65, Figure 22. c – d; 2013: 75). At least 20 of such projectile points interpreted as spearheads have been recovered from the Late Bronze to Iron Age (10th – 6th century BC) hill-forts of Moskenai, Narkunai, Nevieriske, Sokiskiai and Velikuskiai in Lithuania (Grigaviciene 1986a: 68 – 73, Figure 15. 9 – 11, 18. 1 – 4; 1986b: 112, 114, Figure 20. 13 – 18; 1995: 107 – 109, 269, Figure 58, 59. 1; Luik 2013: 26 – 27, Figure 2 – 3; Luik and Maldre 2007: 19 – 20, 24, Figure 26 – 27; Volkaitė-Kulikauskienė 1986: 27 – 28, Figure 32). One specimen from Nevieriske (Grigaviciene 1986a: 70, Figure 15. 9) is provided with a rivet hole at the base. In Estonia two of these weapons have been found in the fortified settlements of Ridala and Aheru, a few have been found in Latvia (Luik and Maldre 2007: 20).

The artefact from Kemnitz fits well into this context from the aspects of raw material choice, size, artefact morphology, find location and chronology and is therefore interpreted as spearhead as well. An interesting aspect is that the finds from Mecklenburg are mainly recovered from rivers and have been interpreted as sacrifices, whereas those found in Brandenburg derive mainly from settlement sites close to the river Havel (Bräunig 2002: 546,

Schoknecht 1983: 47). The spearhead from the settlement site of Kemnitz on the banks of the Havel fits well into the pattern here.

Fractures of the tips of bone projectile points occur typically in case of an impact on a hard material and are one indicator for the use of an artefact as a projectile (Ikäheimo et al. 2004: 15; Petillon 2006: 85–98; Stodiek 1991: 250–254; 2000: 74, 76; Stodiek and Paulsen 1996: 19, 35). The breakage of the point of the Kemnitz implement is old and the location as well as the morphology of the fracture resembles the impact fractures shown by Petillon (2006: 89–93, Figure 53–58) and Stodiek (1991: 253, Figure 7; 2000: 74, Figure 5).

To get some ideas about how these objects were produced an attempt was made to build a replica. Holding a sheep tibia upright in one hand with its distal end on the ground and chopping off flakes of bone with a knife or an axe to get an oblique tip, turned out to be a rather ineffective and inaccurate method. When the tibia was laid down with its caudal side flat on the ground or on a stone anvil instead—which is the only stable position to lay down a sheep tibia due to the *Margo cranialis*—the cutting facet is positioned exactly perpendicular to the *Facies caudalis*. So the general shape can be produced by a single blow using a (metal?) blade held at an oblique angle to the shaft axis (Plate XI. Figure 1). Furthermore, it turned out to be impossible to produce a cut proceeding from the lateral to the medial side on a right tibia when the axe was held in the left hand. If this projectile was produced using this easy method, the craftsmen was probably right-handed.

One 42 mm long fragment from the right tibia of a hare (artefact no. 5, Plate XI. Figure 2) is probably not a tool in the strict sense that it had been worked intentionally to become a tool for a certain purpose. This object displays many sub-parallel scraping or abrasion marks running perpendicular to the long bone axis on each of its three sides (Plate XI. Figure 2. b). What this worked bone was used for remains uncertain. The only similarity coming into my mind here is the irregular distribution of marks on bones used as anvils for the production of metal sickles (Anderson et al. 2014; Gal and Bartosiewicz 2012; Moreno-Garcia et al. 2005). However, the majority of these anvils have been made of metapodiae of cattle and the sickle makers marks look completely different from the traces on the hare bone from Kemnitz. The fragile hare bone also would not have been stout enough for such kind of action. The anvils are similar only insofar as the randomness of the distribution of the marks suggests that the implement rather may have been used as an object to support some kind of work instead of being the subject of an activity itself.

CONCLUSION

The five bone artefacts found within the rescue excavation at the Middle Bronze to Early Iron Age site of Kemnitz, Bramdenburg, Germany, are neither spectacular nor extraordinary. Four of them are simple pointed artefacts for which a multitude of functions is possible, implement no. 5 may not even be a tool in strict sense, but rather an object used to support something. Except for tool no. 2, all artefacts can be classified as class Ⅱ or expedient tools according to Choyke (1997) insofar as raw materials were chosen, which could be transformed into the desired shape with minimum effort (pig fibula, sheep/goat tibia, long bone splinter), a large part of the surface shows manufacturing traces (tool no. 1 and 4) and / or only small parts of the surface show use wear (fibula pin no. 3). The only artefact that maybe called a class I tool is implement no. 2 since its regular rectangular cross-section, the specifically formed tip, the elaborate surface finish and the use-wear allows it to be classified as a "carefully planned tool".

The small number of artefacts does not allow general conclusions about the material culture of the people at Kemnitz, but the finds may help to develop chronological and geographical distribution patterns of certain types of artefacts in combination with similar finds. In particular, the probable spearhead fits well into the context of this find category and adds one find point to the distribution pattern of bone spearheads in Europe. The growing amount of evidence for this find category shows a predominantly circum-Baltic distribution where the Eastern Baltic finds are older (Late Bronze Age) than the Western Baltic finds, which may point to an East to West movement of this weapon type.

Peculiar, but nevertheless interesting, is artefact no. 5. In my experience, bones showing a multitude of nonspecific traces are not so rare in archaeozoological assemblages, but since they usually refuse a clear taphonomic or functional interpretation, they are often either not analysed as artefacts and not published at all or at most shown as peculiar miscellaneous items. A target-oriented search for such artefacts may generate interesting results.

ACKNOWLEDGEMENTS

I would like to thank Daniela Nordholz for the permission to analyse her find material. Special thanks go to Dirk Heinrich, Wolfgang Lage, Hans Frisch and Harm Paulsen of the Archäologisch-Zoologische Arbeitsgruppe Schleswig for support and access to their collection. Wolf-Rüdiger Teegen (ArchaeoBioCenter, Universität München) kindly photographed the artefacts. Andrea Hohlt used a replica on her pottery. Heidi Luik provided me with valuable references. Daniela Nordholz and Lena Wöhlke edited the manuscript, Alice Choyke reviewed the article in press.

REFERENCES

Anderson, P. C., Rodet-Belarbi, I. and Moreno-García, M. (2014) Sickles with Teeth and Bone Anvils. In: A. L. van Gijn, J. C. Whittaker and P. C. Anderson (eds.) *Explaining and Exploring Diversity in*

Agricultural Technology. Early Agricultural Remnants and Technical Heritage 2, 118 – 132. Oxford, Oxbow.

Becker, C. (2001) Bone points—no longer a mystery? Evidence from the Slavic urban fortification of Berlin-Spandau. In: A. M. Choyke and L. Bartosiewicz (eds.) *Crafting Bone: Skeletal Technologies through Time and Space*. British Archaeological Reports International Series 937, 129 – 148. Oxford, Archaeopress.

Becker, C. (2009) Über germanische Rinder, nordatlantische Störe und Grubenhäuser—Wirtschaftweise und Siedlungsstrukturen in Hitzacker-Marwedel. *Beiträge zur Archäozoologie und Prähistorischen Anthropologie* 7, 81 – 96.

Beech, M. (1995) The animal bones from the Hallstatt settlement of Jenstejn, Central Bohemia, Czech Republic. In: D. Dreslerova and M. Beech (eds.) *A Late Hallstatt settlement in Bohemia -Excavations at Jenstejn*, 99 – 140. Prague, Institute of Archaeology.

Bräunig, R. (2002) Verbreitung und Gebrauch von Knochenlanzenspitzen während der vorrömischen Eisenzeit und älteren römischen Kaiserzeit. *Ethnografisch-Archäologische Zeitschrift* 43, 543 – 560.

Buck, D.-W. R. (2002) Schiff ahoi. Bronzezeitliche Keramik mit Bootsdarstellungen von Kemnitz, Landkreis Potsdam-Mittelmark. *Archäologie in Berlin und Brandenburg* 2001, 70 – 73.

Choyke, A. M. (1996) Worked animal bone at the Sarmatian site Gyoma 133. In: A. Vaday (ed.) *Cultural and Landscape Changes in South-East Hungary II : Prehistoric, Roman Barbarian and Late Avar Settlement at Gyoma 133 (Békés county Microregion)*, 307 – 322. Budapest, Archaeolingua.

Choyke, A. M. (1997) The bone tool manufacturing continuum. *Anthropozoologica* 25 – 26, 65 – 72.

Gál, E. and Bartosiewicz, L. (2012) A radiocarbon-dated bone anvil from the chora of Metaponto, southern Italy. *Antiquity Project Gallery* 85 (331).

Grigalaviciene, E. (1986a) Nevieriskes Piliakalnis. *Lietuvos Archeologija* 5, 52 – 88.

Grigalaviciene, E. (1986b) Sokiskiu piliakalnis. *Lietuvos Archeologija* 5, 89 – 138.

Grigalaviciene, E. (1995) *Zalvario ir ankstyvasis gelezies amzius Lietuvoje*, Vilnius, Mokslo ir Enciklopediju Leydykla.

Ikäheimo, J. P., Joona, J.-P. and Hietala, M. (2004) Wretchedly poor, but amazingly practical: Archaeological and experimental evidence on the bone arrowheads of the Fenni. *Acta Borealia* 21(1), 3 – 20.

Kaul, F. (1988) *Da våbnene tav. Hjortespringfundet og dets baggrund*. Copenhagen, Nationalmuseet.

Küchelmann, H. C. (2002) Geschliffen und poliert. Knochenwerkzeuge der jüngeren Bronzezeit aus Kemnitz, Landkreis Potsdam-Mittelmark. *Archäologie in Berlin und Brandenburg* 2001, 73 – 75.

Küchelmann, H. C. (2011) *Tierknochen aus der Siedlung Bentumersiel bei Jemgum, Landkreis Leer (Ostfriesland)*. Report submitted to the Niedersächsisches Institut für historische Küstenforschung Wilhelmshaven. Bremen, Knochenarbeit. online: https://www.knochenarbeit.de/bentumersiel – skn.

Küchelmann, H. C. (2013) Tierknochen aus der Siedlung der Vorr? mischen Eisenzeit und R? mischen Kaiserzeit Bentumersiel bei Jemgum, Ldkr. Leer (Ostfriesland). *Siedlungs – und Küstenforschung im südlichen Nordseegebiet* 36, 63 – 85.

Luik, H. (2006) For hunting or for warfare? Bone arrowheads from the late Bronze Age fortified settlements in Eastern Baltic. *Estonian Journal of Archaeology* 10(2), 132 – 149.

Luik, H. (2013) Late Bronze Age bone crafting in the Eastern Baltic: Standardization of artefact types and individual ingenuity. *Estonian Journal of Archaeology* 17(1), 24 – 37.

Luik, H. and Maldre, L. (2007) Bronze Age bone artefacts from Narkûnai, Nevieriske and Kereliai fortified settlements. Raw materials and manufacturing technology. *Archaeologia Lituana* 8, 5 – 39.

Maldre, L. (2001) Bone and antler artefacts from Otepää Hill-Fort. In: A. M. Choyke and L. Bartosiewicz (eds.). *Crafting Bone: Skeletal Technologies through Time and Space*. British Archaeological Reports International Series 937, 19 – 30. Oxford, Archaeopress.

Moreno-Garcia, M., Esteban Nadal, M., Rodet-Belarbi, I., Pimenta, C., Morales, A. and Ruas, J. P. (2005) *Bone Anvils: Not Worked Bones but Bones for Working*. Poster presented at the meeting of the ICAZ Worked Bone Research Group Veliko Turnovo Bulgaria.

Petillon, J.-M. (2006) *Des Magdaleniens en Armes. Technologie des Armatures de Projectile en Bois de Cervide du Magdalenien superieur de la Grotte d'Isturitz (Pyrenees-Atlantiques)*. Artefacts 10. Treignes, Editions du Cedarc.

Raddatz, K. (1954) Einige Waffen der vorrömischen Eisenzeit aus Norddeutschland. *Offa* 13, 63 – 68.

Schoknecht, U. (1983) Mecklenburgische Knochenlanzenspitzen aus germanischer Zeit. *Jahrbuch Bodendenkmalpflege in Mecklenburg-Vorpommern* 1982, 47 – 66.

Schwarz-Mackensen, G. (1976) *Die Knochennadeln von Haithabu*. Berichte über die Ausgrabungen in Haithabu 9. Neumünster, Wachholtz.

Stodiek, U. (1991) Erste Ergebnisse experimenteller Untersuchungen von Geweihgeschoßspitzen des Magdalénien. In: Fansa, M. (ed.) *Experimentelle Archäologie in Deutschland Bilanz* 1991. Archäologische Mitteilungen aus Nordwestdeutschland Beiheft 6, 245 – 256.

Stodiek, U. (2000) Preliminary results of an experimental investigation of Magdalenian antler points. *Anthropologie et Préhistoire* 111, 70 – 78.

Stodiek, U. and Paulsen, H. (1996) *"Mit dem Pfeil, dem Bogen..."*: Technik der steinzeitlichen Jagd. Archäologische Mitteilungen aus Nordwestdeutschland Beiheft 16, Oldenburg, Isensee.

Teegen, W.-R. (1999) *Vorbericht über die Ausgrabungen von 1996 bis 1999 auf dem mehrperiodigen Fundplatz von Karsdorf, FSt. 9 (Burgenlandkreis)*. unveröffentlichter Bericht, Universität Leipzig.

Volkaitė-Kulikauskienė, R. (1986) Narkunu didziojo piliakalnio tyrinė jimu rezultatai (Apatinis kultūrinis sluoksnis). *Lietuvos Archeologija* 5, 5 – 49.

Maintenance, Inheritance and Memory

V-perforated Ivory Buttons from the Los Castillejos Chalcolithic Site in Las Peñas de los Gitanos (Granada, Spain)

Manuel Altamirano García

GEA Research Group. Department of Prehistory and Archaeology. Faculty of Philosophy and Letters, University of Granada, Spain

Abstract: The aim of this paper is to show some special objects from the past that were carefully manufactured, curated and passed down from hand to hand through generations. An assemblage of nine V-perforated buttons dating to the Ⅲ milennium BC has been studied here from perspective of their individual biographies. The selection of raw material from which these items were made was extremelly careful, the main source being elephant ivory, making them really valuable, prestigious and objects for representation. This research focuses on the traceological analysis of the objects' surfaces using binocular and ESEM microscopy which show that these artifacts have very smooth and worn surfaces. Moreover, these buttons were repaired (curated) repeatedly with the clear intention of keeping them in use over long periods of time. All these data, together with the information from the archaeological record and some ethnographical parallels, are indispensable for understanding and explaining how and why these objects were manufactured, used, curated and discarded in the past.

Keywords: V-perforated buttons, manufacture, object biography, Los Castillejos de Montefrío, Copper Age

CONTEXT AND CHRONOLOGY

The archaeological site of Los Castillejos in Las Peñas de los Gitanos is one of the most important settlements in the study and understanding of social developments during the late prehistory in the central part of southern Spain. The site is located in the so-called Los Montes region, on the north-western edge of the Granada Basin.

Although the main focus here will be on the late prehistoric settlement of Los Castillejos, other sites have been identified in the Las Peñas de los Gitanos site including a megalithic necropolis (linked to the Chalcolithic occupation of the prehistoric site) as well as a pre-Roman village and both Roman and medieval remains.

Known since the second half of the nineteenth century (Góngora 1868), the Los Castillejos site was partially and occasionally studied (Gómez 1907, 1949; Mergelina 1941-42, 1945-46; Tarradell 1947, 1952). It was, nevertheless, not systematically excavated until 1971, when the Department of Prehistory and Archeology of the University of Granada undertook several seasons of fieldwork on the site (Arribas 1976; Arribas and Molina 1977, 1978, 1979a, 1979b). Both new excavations and restoration works were carried out some years later by archaeologists at this department within the framework of an ambitious research project (Ramos et al. 1997; Afonso and Ramos 2005; Cámara et al. 2010).

Thanks to all this fieldworkand based on the recent radiocarbon dates (AMS), the broad occupational sequence has been organized into 30 phases and sub-phases(Afonso et al. 1996; Cámara et al. 2005; Nachasova et al. 2007; Cámara et al. 2010). The prehistoric occupation of Los Castillejos begins in the Early Neolithic to the Early Bronze Age, although here the main focus will be on special objects from the Copper Age, which here includes the so-called Period V to VIII and Phase 16b to 23c (Table 1).

Table 1. Chronology of the Chalcolithic phases from Los Castillejos.

PERIOD			CHRONOLOGY	PHASE
Early Copper Age		V	3300–3000 BC	16b, 17
Middle Copper Age		VI	3000–2600 BC	18, 19
Late Copper Age	Tardio CA	VII	2600–2400 BC	20, 21, 22
	Final CA	VIII	2400–2000 BC	23a, 23b, 23c

Regarding the transformation and working of hard animal materials, it has become clear that this represented an important craft on the site throughout the third millennium BC, with evidence for some types that the manufacturing ethos wasdeeply influenced by strong Neolithic manufacturing and cultural traditions (Salvatierra 1982; Altamirano 2013, 2014b, 2014c). An assemblage of 155 osseous tools and ornaments from the Copper Age layers came to light during excavations carried out in 1947, 1971 and 1974 (Tarradell1947; Arribas 1976; Arribas and Molina 1977, 1978, 1979a, 1979b), with a clear majority of objects being made from mammal bones (85%). Others types of raw materials such as red deer antler, marine mollusk shells and elephant ivory were also important raw materials for manufacturing objects, although clearly and proportionally much less significant than bone (Altamirano 2013, 2014a, 2014b, 2014d).

This paper mainly focuses on a small assemblage of osseous ornaments whose special features make them unique objects, appropiate for a separate study. They comprise nine V-perforated buttons documented in the Middle and Late Copper Age deposits of the Los Castillejos site and were discovered within the habitation area, being notable for the two special raw material used in their manufacture: ivory and red deer antler.

The so-called Bell Beaker "package" has traditionally been characterized by certain metal and ceramic objects, with less attention paid to other items made from other raw materials such as bone. It is important to highlight the presence of V-perforated buttons, normally made from bone, ivory and more rarely, red deer antler. Based on the archaeological record from Europe and the Iberian peninsula, as Bell Beaker groups developed, there would have been an increased demand for this prestigious material.

This type of artifact emerged in the archaeological record of Iberia during the Early Copper Age with a large increase in the number of such V-perforated buttons from the III millenium BC contexts (Harrison and Gilman 1977; Espadas et al. 1987; Liesau 2016), although many other assemblages of V-perforated buttons have also been discovered in Early, Middle and Late Bronze Age contexts in Iberia, frequently found within funerary context as grave goods (Fonseca 1988; Uscatescu 1992; Altamirano 2013, 2018; López 2006a, 2006b).

Based on the available archaeogical data from Los Castillejos, all of these V-perforated buttons appeared before the Late Middle Copper Age (*circa* 2800 cal BC). One the one hand, four of these artifacts date to the Middle Copper Age. This corresponds to the Period VI (3000 – 2600 cal BC), with deposits formed from dwellings with walls made from mud and other organic materials, the presence of the first Bell Beaker ceramics (the so-called Maritime Style), some archer's wrist guards bracelets, copper projectile points of the "Palmela Type" and some others artifacts such as long, curved clay cylinders (Arribas and Molina 1978, 1979a; Cámara et al 2010).

On the other hand, three V-perforated buttons came to light within the Late Copper Age levels (*Tardio* and Final CA). This corresponds to Periods VII (2600 – 2400 cal BC) and VIII (2400 – 2000 cal BC), when many more permanent dwellings with stone foundations and wattle and daub walls were built. Bell Beaker pottery (the so-called Ciempozuelos Style) and the production of metal objects expands to a notable extent in this period (Arribas and Molina 1978, 1979a, 1979b; Cámara et al 2010).

Finally, two more buttons were found in the deposits dating to a transitional period between the Middle and Late (*Tardio*) Copper Age (Table 1).

RAW MATERIAL AND MANUFACTURE

The choice of raw material from which the objects were manufactured needs to be considered first. This is certainly something that, although *a priori* may seem trivial, is very significant both technically and culturally. This choice may be related to various technical and social factors (Choyke and Bartosiewicz 2009), such as the availability of the raw material, the physical and mechanical properties which sets the limits for what the given raw material could be used to produce, and, finally, the possible significance or attitude towards a particular species or a specific part of the animal of both the craftspeople manufacturing these objects and the individuals using V-perforated buttons.

Based on our research, it seems most of the tools from the Los Castillejos site that were made from split mammalian long bones were linked to the domestic sphere. Such long bones were easily obtained from local caprinae and cattle. On the other hand, especial objects such as ornaments, were normally made from less common or even what may have been considered exotic raw materials. This is the situation raised up by the raw material analysis regarding the V-perforated buttons assemblage. Except for one specimen made from red deer antler, the remainder of these buttons were made from elephant ivory. These are materials that might well be considered special, even exotic and extremely valuable, especially in the case of elephant ivory (Plate XII. a – c & e – h; Plate XIII).

The shape of V-perforated button made from red deer antler (Plate XII. d), is less usual (the so-called *Tortoise shell* type). Some parallel pieces have been documented at other contemporary sites, although these were made either from bone or ivory (Enríquez 1982; Uscatescu 1992; Maicas 2007; Pau 2007; Liesau 2016). The specimen from Los Castillejos was made from an elongated segment extracted from the cortical material of either beam segment "A" or "B" based on the traceological analysis and its slightly curved profile. No evidence remains regarding the reduction (débitage) proccess to obtain the blank. However, given the morphological features of this artifact, extraction seems to have been the method used to get the blank from which this button was manufactured.

The ivory objects were all made from elephant tusk, indicated by the presence of the *Schreger lines* on their surfaces (Banerjee and Huth 2012), although their presence cannot always be observed (Krzyszkowska 1990). The possible origin of this raw material is not clear at all. According to Dr. Thomas X. Schuhmacher, at least one of these buttons may have been manufactured on a section of fossil ivory (*Palaeoloxodon antiquus*) based on its appearance, structure and material features. The use of fossil ivory has also been documented elsewhere on contemporary sites in the Iberian Peninsula such as the Camino de las Yeseras site (Liesau 2016), although mainly used to manufacture small objects (Liesau and Moreno 2012).

Recent research has claimed that most of artifacts made from ivory dating to the third millennium appear to have been made from Asian elephant tusk instead of tusks of African elephant. This tendency changed in part from the end of the third and the begining of the second millennium, when African elephant ivory notably increased (Banerjee et al. 2012; Schuhmacher 2012). However, research carried out by Sonia O'Connor has highlighted how much controversial these analysis are, being DNA the only method to tell apart both African and Asian ivories with confidence, having carried out other types chemical techniques as well as vibrational spectroscopy to measure Schreger angle (O'Connor et al. 2011). She points out that most results might not be consistent enough, because from her experience they would be mainly affected by the age, sex and diet of the individual elephants rather than by their species.

Based on the research mentioned above, it would be necessary to carry out several future analyses in order to confirm whether their raw material has either an Asian or African origin regarding this ivory button assemblage from Los Castillejos.

The traceological analysis has shown that these V-perforated buttons from the Los Castillejos site were carefully made. Their manufacture was planned and carried out using the same techniques and tools. Although the surface of these artifacts is mostly worn and smoothed due to intensive use, several manufacturing traces have been identified thanks to a microscopical analysis using a binocular microscope as well as ESEM.

By and large, three strictly technical operations are involved in the manufacture of V-perforated ivory buttons: cutting (longitudinal or transversal), surface abrasion and a converging double perforation on the under surface.

In one case, it is also possible to observe the inner structure of the elephant tusk with its characteristic concentric Schreger lines. Thus, we know that the tusks were divided into several thick and transverse slices that were subsequently transformed into artifacts, such as the great prismatic button from Los Castillejos (Plate XIII). As some other artifacts suggest, blanks were also obtained by extracting elongated prismatic rods from which smaller portions were cut in order to manufacture the buttons (Mérida 1997). Although no direct evidence has been observed, the use of metal tools to process the tusk and obtain the blanks is at least a possibility, as recent research has noted, based on the found material from the pre-Bell Beaker site of Valencina de la Concepcion, in Seville (Garcia Sanjuan et al. 2013; Luciañez 2018). Based on the archaeological record, the presence of worked elephant ivory (not fossil ivory) in the south of the Iberian Peninsula, seems to occur together with the first metal tools such as awls and small saws (Plate XIV), between the end of the fourth and the begining of the third millennia BC (Schuhmacher 2012; Nocete et al. 2013).

In all cases, preforms were modified by abrading their surfaces to create the desired shape of the object (hemispheric, prismatic, pyramidal, etc.). Finally, a characteristic technique used in manufacturing the buttons is the V-perforation, an aesthetic element which allows the entire surface of each piece to be shown without revealing the form of attachment. The perforations on the ivory buttons display a cylindrical or slightly conical profile. They were probably drilled using a thin metal awl with a turning movement, possibly with the help of a bow or pump system drill (Mérida 1997).

It is also noteworthy that at least in one case the surface of the button was decorated. The big prismatic button (2400 – 2000 cal BC), displays series of parallel, straight lines from its top to its base over one of its sides. These incised lines were probably made using a thin copper awl, based on their narrow, V-shaped profile.

USE, CURATION AND INHERITANCE

The V-perforated button assemblage is associated with the various Bell Beaker levels at the Los Castillejos site. Several of these artifacts are very well preserved and display a very heavy degree of use, evidenced on a macro-scopic level by the intense brightness over their surfaces (Plate XV). Furthermore, in four cases, the V-perforation buttons needed to be repaired (curated), sometimes more than once (Plate XIV).

The real function of objects and how they were really used in the past is still unkown. Commonly called buttons, they have traditionally been thought to be a type of dress accesory that might have been sewn to clothing as real buttons or simple rich ornaments (Barciela 2006, 2012; Liesau 2016).

Although some of these objects could have been used as buttons, given their shape and measurements, others may have been used as pendants (even part of a complex necklace), rather than having been sewn onto clothes. Based on experimental studies, the narrow and convergent V-perforation would not have allowed a needle to pass through it (Ugas 1982).

Traceological analysis will not tell us with any certainty how the buttons from Los Castillejos were used in the past. However, the extremely worn and smooth surfaces (less marked on their inner face) of these buttons suggests that they were sewn onto clothes against which the button consistently rubbed. Whether they were to be carried and shown or not, these buttons were certainly meant to be worn and displayed and their main function and mission were to act as real symbolic elements reflecting and acting on social and cultural rules and beliefs.

The high degree of use wear evidenced by the totally worn surfaces of the V-perforated buttons, together with the special raw material with which they were manufactured and the evidence for repeated curation, provide an approximate idea of how precious these objects were for people in the past. Based on all this evidence, it is suggested here that some of these buttons may have been handed down over generations and served as important markers of certain kinds (they may have represented multiple identities as in gender, age cohort, identity of place, or social rank) of social identity connected to Bell Beaker society beyond the boundaries of this particular settlement. The inheritance of these and other objects would have made them extremely significant and valuable to their owners, making them a materialization of the memory and meaning of the ancestors to the living (Choyke 2001, 2006).

DISCUSSION

Raw materials represent a very valuable source of information about the objects that could be manufactured from hard osseous materials from animals. Based on this research, exotic and even precious materials were especially selected at times to manufacture certain types of ornaments. The analysed buttons discussed here were made from elephant ivory (except for one find made from red deer antler), a foreign, exotic material that was systematically used to produce itmes that were particularly important and socially valuable (López 2011; Altamirano 2013; Liesau 2016).

Many researchers have generally assumed that attributes such as shape, hardness, brittleness or, more or less, easy availability of raw material, were important criteria in deciding to manufacture an object in ivory, bone or antler. Most likely, these criteria are indeed important, since it is true that certain materials can be worked more easily or are simply more available than others in particular contexts. However, it must be considered whether it was only these sorts of questions that conditioned the choice of a particular raw material. Some of these materials were certainly characterized by a certain symbolism, beliefs, metaphor or charged social values (necessary in representation and display) of a specific group or community. As Corina Liesau points out (Liesau 2016), these V-perforated buttons, among other items, would reveal the elitist characteristic of grave goods because they are normally found inside funerary contexts which also contain some other objects made from gold, copper or cinnabar.

It must be remembered that people living in a particular cultural milieu may have attributed various socially-weighted attributes to certain animals, especially to those with more direct economic or subsistance impact either in the domestic or in wild sphere. These animals may be considered ancestral, protective totems, so that objects made from their bones may also have special apotrapaic powers as amulets (Choyke 2010; Choyke and Kovats 2010; van Gijn 2017).

Human life is inextricably associated with the material world. Portable objects such as clothing, ornaments, food or tools are fully involved and integrated in many kinds of everyday activities. At the same time, people use objects, both consciously and unconsciously to transmit information about themselves and their place(s) within both their narrow social unit and outside it to various target audiences of different scales (Appadurai 1986; Hodder 2012).

The biography of objects, therefore, may be intimately linked to the history of people. When used or carried for a long time in the domestic sphere or within the urban settlement in general, or during travels away from home for trading or war missions, these special items tend to be closely associated with the memory of the person or people who were specially close to them (Choyke and Daroczi-Szabó 2010: 242). At the same time, as object for display, V-perforated buttons had a far-flung audience who read their social message in Bell Beaker times and ivory may well have been recognized as a high status material beyond Iberia.

Thus, we might speak in some cases of what have been called heirlooms, special items, such as ornaments or other symbolic objects that are used over a long period of time, curated and may be transmitted from person to person or group to group over several generations (Choyke 2009).

From this perspective, the same object can transmit different messages about both individual and group identities (age, sex, class, locational, religious, etc.) at the same time (Weissner 1983). The maintenance (curation) and continued use of certain objects may have reflected back on ancestral beliefs or memories and traditions, perhaps in an effort to legitimize status as well as other aspects of social identity (Ashby 2006, 2011: 11).

The decorative motifs used on various objects, such as personal ornaments, may seem similar from our current perspective. In the past, however, even slight differences may have had important meaning. These symbols might have been true markers of identity for those able to read the material language and meaning implicit in their signs. Hence, social context is important although its details have largely been lost to researchers (Ashby 2011).

The nine V-perforated buttons studied here display extremely worn and smooth surfaces resulting from having been in circulation over long periods of time. The especial raw material they were made from, the care and effort put into their manufacture and the high degree of use observed on their worn surfaces all show how precious these objects were for their owners.

It is increasingly clear that both osseous tools and ornaments, compared with most ceramic artifacts or textiles and leather, may easily outlast their owners, the settlement to they were linked and be transmitted from generation to generation. They would have acted as links between the living and the dead in burial ritual and transmitted messages about the people wearing them in a local context and in far away situations as well. Thus, certain marker objects such as the Bell Beaker V-perforated buttons became important and meaningful to the people who used them as sources of social authentification and social value within various contexts of both personal and family identities (Choyke 2001; 2006, 2009; Choyke and Daroczi-Szabó 2010: 245) and long-distance contacts as well.

REFERENCES

Afonso, J. A., Molina, F., Cámara, J. A., Moreno, M., Ramos, R. and Rodríguez, M. O. (1996) Espacio y tiempo. La secuencia en Los Castillejos de Las Peñas de Los Gitanos (Montefrío, Granada). In: J. Bosch and M. Molist, (eds.) *I Congrés del Neolític a la Península Ibérica. Formació e implantació de les comunitats agrícoles (Gavà-Bellaterra, 1995). Actes. Vol. 1. Rubricatum* 1(1), 297–304.

Altamirano, M. (2013) *Hueso, asta, marfil y concha: aspectos tecnológicos y socioculturales durante el III y II milenio A. C. en el sur de la Península Ibérica.* Unpublished thesis, Department of Prehistory and Archaeology, University of Granada.

Altamirano, M. (2014a) Los peines óseos de Los Castillejos en las Peñas de los Gitanos (Montefrío, Granada), Movilidad Contacto y Cambio. *Actas del II Congreso de Prehistoria de Andalucía*, 361–369.

Altamirano, M. (2014b) Hueso, asta y marfil: manufactura de artefactos durante el III milenio AC en el poblado de Los Castillejos (Montefrío, Granada). *Saguntum* (*P. L. A. V.*) 46, 21 – 40.

Altamirano, M. (2014c) Not only bones. Hard animal tissues as a source of raw material in 3rd millennium BC South-Eastern Iberia. *Menga. Revista de Prehistoria de Andalucía* 5, 43 – 67.

Altamirano, M. (2014d) Uso y mantenimiento de objetos. Botones y peines de marfil, hueso y asta de ciervo de Los Castillejos de Montefrío (Granada). *Antiquitas* 26, 155 – 160.

Altamirano, M. and Alarcón, E. (2018) Bone tools for the deceased: Approaches to the worked osseous assemblage from the Bronze Age funerary cave of Biniadris (Menorca, Spain). *Quaternary International* 472 (2018), 108 – 114.

Arribas, A. (1976) Las bases actuales para el estudio del Eneolítico y la Edad del Bronce en el Sudeste de la Península Ibérica. *Cuadernos de Prehistoria y Arqueología de la Universidad de Granada* 1: 139 – 155.

Arribas, A. and Molina, F. (1977) El poblado de Los Castillejos en Las Peñas de los Gitanos (Montefrío, Granada). Campañas de excavaciones de 1971 y 1974. *XIV Congreso Nacional de Arqueología* (*Vitoria, 1975*), 389 – 406, Zaragoza.

Arribas, A. and Molina, F. (1979a) El poblado de los Castillejos en las Peñas de los Gitanos (Montefrío, Granada): campaña de excavaciones de 1971: el corte no 1. *Cuadernos de Prehistoria de la Universidad de Granada*. Serie Monográfica 3. Granada.

Arribas, A. and Molina, F. (1979b) Nuevas aportaciones al inicio de la metalurgia en la Península Ibérica. El poblado de Los Castillejos de Montefrío (Granada). In M. Ryan (ed.) *The Origins of Metallurgy in Atlantic Europe*. Proceedings of the Fifth Atlantic Colloquium, Dublín, 30th March to 4th April 1978, 7 – 34.

Appadurai, A. (1986) Introduction: Commodities and the politics of value. In: A. Appadurai (ed.) *The Social Life of Things: Commodities in Cultural Perspective*, 10 – 19.

Ashby, S. P. (2006) *Time, Trade and Identity: Bone and Antler Combs in Northern Britain c. AD 700 – 1400*, Unpublished thesis, Department of Archaeology, University of York.

Ashby, S. P. (2011) An Atlas of Medieval Combs from Northern Europe, *Internet Archaeology* 30.

Banerjee, A. and Huth, J. (2012) Investigation of Archaeological Ivory. In: A. Banerjee, J. A. Lopez and T. X. Schuhmacher (eds.) *Marfil y elefantes en la Península Ibérica y el Mediterráneo occidental*. Actas del coloquio internacional en Alicante el 26 y 27 de noviembre de 2008. Iberia Archaeologica 16, 15 – 28. Deutsches Archäologisches Institut. Diputación de Alicante. MARQ, Museo Arqueológico de Alicante.

Barciela, V. (2006) *Los elementos de adorno de El Cerro de El Cuchillo* (*Almansa, Albacete*). Diputación de Albacete (Memorias del Instituto de Estudios Albacetenses, 172), Albacete.

Barciela, V. (2012) Tecnologíadelmarfilenla Edaddel Broncedela Meseta Sur (España). In: A. Banerjee, J. A. Lopez and T. X. Schuhmacher (eds.) *Marfil y elefantes en la Península Ibérica y el Mediterráneo occidental*. Actas del coloquio internacional en Alicante el 26 y 27 denoviembre de 2008. Iberia Archaeologica 16, 199 – 214. Deutsches Archäologisches Institut. Diputaciónde Alicante. MARQ, Museo Arqueológicode Alicante.

Cámara, J. A., Molina, F. and Afonso, J. A. (2005) La cronología absoluta de Los Castillejos en Las Peñas de los Gitanos (Montefrío, Granada). *Actas del III Congreso del Neolítico en la Península Ibérica* (*Santander, 5 a 8 de octubre de 2003*), 841 – 852.

Cámara, J. A., Afonso, J. A. and Molina, F. (2010) *La ocupación de las Peñas de los Gitanos* (*Montefrío, Granada*) *desde el Neolítico al mundo romano. Asentamiento y ritual funerario*. Ayuntamiento de Montefrío,

Granada.

Choyke, A. M. (2001) A quantitative approach to the concept of quality in prehistoric bone manufacturing. In: H. Buitenhuis and W. Prummel (eds.) *Animals and Man in the Past*, ARC-Publicatie 41, 59 – 66. Groningen, The Netherlands.

Choyke, A. M. (2006) Bone tools for a lifetime: experience and belonging. In: Astruc, L., Bon, F., Léa, V., P. Y. Milcent and S. Philibert (eds.) *Normes techniques et practiques sociales. De la simplicité des outillages pré-et protohistoriques*. XXVI rencontres internationales d'archéologie et d'histoire d'Antibes, 49 – 60.

Choyke, A. M. (2009) Grandmother's awl: Individual and collective memory through material culture. In: I. Barbiera, A. M. Choyke and J. Rasson (eds.) *Materializing Memory: Archaeological Material Culture and the Semantics of the Past*, 21 – 40.

Choyke, A. M. (2010) The bone is the beast: Animal amulets and ornaments in power and magic. In: D. Campana, P. Crabtree, S. D. DeFrance, J. Lev-Tov and A. Choyke (eds.) *Anthropological Approaches to Zooarchaeology: Colonialism, Complexity and Animal Transformation*, 197 – 209. Oxbow Books: Oxford.

Choyke, A. M. and Bartosiewicz, L. (2009) Telltale tools from a tell: Bone and antler manufacturing at Bronze Age Jászdózsa-Kápolnahalom. *Tiscium* XX, 357 – 376.

Choyke, A. M. and Kováts, I. (2010) Tracing the personal through generations: Late medieval and ottoman combs. In: A. G. Pluskowski, G. K. Kunst, M. Kucera, M. Bietak and I. Hein (eds.) *Bestial Mirrors. Using Animals to Construct Human Identities in Medieval Europe*, 115 – 127. Vienna Institute for Archaeological Science.

Choyke, A. M. and Darócz-Szabó, M. (2010) The complete and usable tool: Some life histories of prehistoric bone tools in Hungary. In: A. Legrand-Pineau, I. Sidéra, N. Buc, E. David and V. Scheinsohn (eds.) *Ancient and Modern Bone Artefacts from America to Russia. Cultural, Technological and Functional Signature*. BAR International Series 2136, 235 – 248.

Enríquez, J. J. (1982) Objetos de adorno personal de la Prehistoria de Navarra. *Trabajos de Arqueología Navarra* 3, 157 – 202.

Espadas, J., Poyato, C. and Caballero, A. (1987) Memoria preliminar de las excavaciones del yacimiento calcolítico de El Castellón, Villanueva de los Infantes. *Oretum* III, 39 – 78. Ciudad Real, Servicio de Publicaciones e Intercambio Científico.

García, L., Luciañez, M., Schuhmacher, T., Wheatley, D. W. and A. Banerjee (2013) Ivory craftsmanship, trade and social significance in the southern Iberian Copper Age: The evidence from the PP4-Montelirio sector of Valencina de la Concepción (Seville, Spain). *European Journal of Archaeology* 16(4), 610 – 635.

van Gijn, A. (2017) Bead Biographies from Neolithic Burial Contexts: Contributions from the Microscope. In: D. Mayer Bar-Yosef and A. Choyke (eds.), *Not Just for Show* 2017, 103 – 114. Oxford, Oxfow Books.

Gómez, M. (1907) *Monumentos arquitectónicos de España*. Madrid.

Gómez, M. (1949) Monumentos arquitectónicos de la provincia de Granada. Misceláneas. *Historia-Arte-Arqueología, Primera Serie, La Antigüedad*, 347 – 390.

Góngora, M. (1868) *Antigüedades prehistóricas de Andalucía, monumentos, inscripciones, armas, utensilios y otros importantes objetos pertenecientes a los tiempos más remotos de su población*. Madrid.

Harrison, R. andGilman, A. (1977) Trade in the second and third Millennia B. C. between Magreb and

Iberia. Ancient Europe and the Mediterranean. Studies in Honour of Hugo Hencker, 91 – 104. Warminster.

Hodder, I. (2012) *Entangled: An Archaeology of the Relationships between Humans and Things.* Wiley-Blackwell.

Hoskins, J. (1998) *Biographical Objects. How Things Tell the Stories of People's Lives.* London, Routledge.

Jones, S. (1999) Historical categories and the praxis of identity: The interpretation of ethnicity in historical archaeology. In: P. P. Funary, M. Hall and S. Jones (eds.) *Historical Archaeology: Back from the Edge*, 219 – 232. London, Routledge.

KrzyszKoWsKa, O. (1990) *Ivory and Related Materials. An Illustrated Guide.* Classical Handbook 3, Bulletin supplement 59. London, Institute of Classical Studies.

Leisner, G. and Leisner V. (1943) *Die Megalithgräber der Iberische Halbinsel. Der Western.* Deustsches Archäelogisches Institut. Abteilung Madrid.

Liesau, C. (2016)Some pretige godos as evidence of interregional interactions in the funerary practices of the Bell Beaker groups of Central Iberia. In: E. Guerra and C. Liesau (eds.) *Analysis of the Economic Foundations Supporting the Social Supremacy of the Beaker Groups.* Proceedings of the UISPP World Congress, Vol. 6, Session B36, 69 – 94. Oxford, Archaeopress Archaeology.

Liesau, C. and Moreno, E. (2012) Marfiles campaniformes de El Camino de las Yeseras (San Fernando de Henares, Madrid). In: A. Banerjee, J. A. López and T. X. Schuhmacher (eds.): *Elfenbeinstudien. Marfil y elefantes en la Península Ibérica y el Mediterráneo occidental.* Actas del coloquio internacional en Alicante el 26 y 27 de noviembre de 2008. Iberia Archaeologica 16, 83 – 94. Deutsches Archäologisches Institut. Diputación de Alicante. MARQ, Museo Arqueológico de Alicante.

López, J. A. (2006a) Distribución territorial y consumo de botones de perforación en "V" en el ámbito argárico. *Trabajos de Prehistoria* 63, 2, 93 – 116.

López, J. A. (2006b) Marfil, oro, botones y adornos en el área oriental del País de El Argar. *MARQ. Arqueología y Museos*1, 25 – 48.

López, J. A. (2011) *Asta, hueso y marfil : artefactos óseos de la Edad del Bronce en el levante y Sureste de la Península ibérica (c. 2500 – c. 1300 cal BC).* Serie Mayor 9, Museo Arqueológico de Alicante (MARQ).

Luciañez, M. (forthcoming) *El marfil en la Edad del Cobre de la península Ibérica. Una aproximación tecnológica, experimental y contextual a las colecciones ebúrneas del mega-sitio de Valencina de la Concepción-Castilleja de Guzmán (Sevilla).* Unpublished thesis, University of Seville.

Maicas, R. (2007) *Industria ósea y funcionalidad: Neolítico y Calcolítico en la cuenca de Vera.* Bibliotheca Praehistorica Hispana. CSIC.

Mérida, V. (1997) Manufacturing process of V-Perforated ivory buttons. In: L. A. Hannus, L. Rossum and R. P. Winham (eds.) *Proceedings of the 1993 Bone Modification Conference, Hot Springs, South Dakota.* Occasional Publication 1, 1 – 11. Archeology Laboratory, Augustana College.

Mergelina, C. (1941 – 1942) La estación arqueológica de Montefrío (Granada) I. Los dólmenes. *Boletín del Seminario de Estudios de Arte y Arqueología* VIII (B. S. A. A.), 33 – 106. Valladolid.

Mergelina. C. (1945 – 1946) La estación arqueológica de Montefrío (Granada) II. La acrópolis de Guirrete (Los Castillejos). *Boletín del Seminario de Estudios de Arte y Arqueología* XII, 15 – 26. Valladolid.

Moreno, M. A. (1982) Los materiales arqueológicos del poblado de Los Castillejos y Cueva Alta (Montefrío) procedente de las excavaciones de 1946 y 1947. *Cuadernos de Prehistoria de la Universidad de Granada* 7, 235 – 266.

Nachasova, I. E., Burakov, K. S., Molina, F. and Cámara, J. A. (2007) Archaeomagnetic Study of Ceramics from the Neolithic Los Castillejos Multilayer Monument (Montefrio, Spain). *Izvestiva*, *Physics of the Solid Earth* 43(2), 170–176.

Nocete, F., Vargas, J. M., Schuhmacher, T. X., Banerjee, A. and Dindorf, W. (2013) The ivory workshop of Valencina de la Concepción (Seville, Spain) and the identification of ivory from Asian elephant on the Iberian Peninsula in the first half of the 3rd millennium BC. *Journal of Archaeological Science* 40, 1579–1592.

Pau, C. (2007) Elementos de adorno personal en época Campaniforme en Sicilia, Cerdeña y Córcega, *Arqueología y Territorio* 4, 23–46.

O'Connor, S, Edwards, H. G. M. and Ali, E. (2011): An interim investigation of the potential of vibrational spectroscopy for the dating of cultural objects in ivory. *ArcheoSciences* (*revue d'archéométrie*) 35, 159–165.

Ramos, U., Afonso, J. A., Cámara, J. A., Molina, F. and Moreno, M. (1997) Trabajos de acondicionamiento y estudio científico en el yacimiento de Los Castillejos en Las Peñas de los Gitanos (Montefrío, Granada). *Anuario Arqueológico de Andalucía* 1993 III, 246–252. Sevilla.

Schuhmacher, T. X. (2012) El marfil en España desde el Calcolítico al Bronce Antiguo. Resultados de un proyecto de investigación interdisciplinar. In: A. Banerjee, J. A. López and T. X. Schuhmacher (eds.) *Elfenbeinstudien. Marfil y elefantes en la Península Ibérica y el Mediterráneo occidental*. Actas del coloquio internacional en Alicante el 26 y 27 de noviembre de 2008. Iberia Archaeologica 16, 45–68. Deutsches Archäologisches Institut. Diputación de Alicante. MARQ, Museo Arqueológico de Alicante.

Tarradell, M. (1947) Un yacimiento de la primera Edad del Bronce en Montefrío, Granada. Avance de los resultados de las últimas excavaciones efectuadas en Las Peñas de los Gitanos. *Crónica del III C. A. S. E.* 52. Murcia.

Tarradell, M. (1952) La Edad del Bronce en Montefrío (Granada). Resultados de las excavaciones en yacimientos de las Peñas de los Gitanos. *Ampurias* XIV, 49–80.

Uscatescu, A. (1992) *Los botones de perforación en "V" en la Península Ibérica y las Baleares durante la Edad de los Metales*. Ed. Foro. Madrid.

Wiessner, P. (1983) Style and social information in the Kalahari San projectile points. *American Antiquity* 48 (2), 253–276.

Bronze Age Faunal Remains from Late Bronze Age Features at Razdolnoe, Eastern Ukraine

Pam J. Crabtree

Center for the Study of Human Origins, Department of Anthropology,
New York University, USA

Abstract: During the summer of 2010, a joint Ukrainian and American team excavated a number of Neolithic and Chalcolithic features at the site of Razdolnoe in eastern Ukraine. An earlier Ukrainian team has already excavated the overlying Late Bronze Age features, but the fauna from these features had never been studied. As part of the 2010 campaign, I identified these animal bones. While the remains were primarily those of domestic animals, including cattle, horses, sheep, and pigs, the remains of a smaller number of wild mammals were also recovered, including red deer, roe deer, and possible aurochs. Although the assemblage is small, it can contribute to our understanding of hunting and herding in the Late Bronze Age of Eastern Ukraine.

Keywords: Eastern Ukraine, Razdolnoe, Late Bronze Age, animal husbandry, hunting

INTRODUCTION

During the summer of 2010, a joint Ukrainian and American team excavated Chalcolithic and Neolithic features at the site of Razdolnoe in eastern Ukraine. The results of these excavations have recently been published, and they are part of a broader ongoing study of the beginnings of animal domestication and pastoralism in eastern Ukraine (see Kotova et al. 2017). Many Chalcolithic sites in eastern Ukraine are covered by extensive remains of later Late Bronze Age occupation. The Chalcolithic levels at Razndolnoe were accessible because an earlier Ukrainian excavation team had removed the Late Bronze Age deposits at the site. However, the fauna recovered from these excavations had never been identified. As part of the 2010 excavation season I identified these faunal remains. Relatively few Late Bronze Age faunal collections from eastern Ukraine have been published (see Morales Muñiz and Antipina 2003 for a summary of the available data). The animal bone data from Razdolnoe can provide some basic information on animal husbandry and hunting practices in Late Bronze Age eastern Ukraine. This study is necessarily a preliminary one, as it was carried out in a field lab without access to a systematic comparative collection. This report will present a brief overview of the faunal remains recovered from Late Bronze Age Razdolnoe and the methods that were used in their identification and analysis.

METHODS

Since I did not have access to a systematic comparative collection, I relied on standard identification manuals (e.g. Schmid 1972) and my four decades of experience in the study of animal bones from prehistoric and early historic sites in Europe and the ancient Near East. All the bone identifications were recorded on coding sheets and then entered into a specialized database manager, FAUNA. FAUNA is designed to record data on archaeological context, animal species, body part, portion, degree of fragmentation, and data on bone measurements, ages at death, and butchery traces (Campana 2010). Animal bone fragments that could not be identified to species were identified to higher order taxa including small artiodactyl ("sheep-sized"), large artiodactyl ("cattle-sized"), and large ungulate (cattle, red deer, or horse). Bone measurements were recorded following the recommendations of von den Driesch (1976), and ageing data were recorded based on both epiphyseal fusion of the long bones (Silver 1969) and dental eruption and wear (Payne 1973; Grant 1982).

ANIMAL SPECIES IDENTIFIED

The animal species identified from the Razdolnoe excavation are listed in Table 1a, and the body part distributions for cattle, caprines, pigs, and horses are shown in Table 1b. The species identified included domestic cattle, sheep, pig, horse, and dog, as well as a small number of wild forest forms, including aurochs (*Bos primigenius*), red deer (*Cervus elaphus*) and roe deer (*Capreolus capreolus*). Aurochs bones were distinguished from domestic cattle on the basis of size. Measurement ranges for wild cattle were published by Bökönyi (1995) and summarized by Bartosiewicz et al. (2006: 32, Table 3). Measurements on the astragalus confirmed the presence of both wild and domestic cattle (Table 2). Measurements taken on the Razdoenoe metatarsi, tibiae, and radii are all below the range of aurochs measurements published by Bökönyi (1995). The measurement data indicate that the Razdolnoe assemblage includes a large number of domestic cattle bones and a small number of aurochs bones.

The domestic animal remains included a small number of bones that could be definitively identified as sheep and a larger number that were simply identified a sheep/goat. No goat bones were identified. For quantitative purposes, the sheep and sheep/goat remains have been combined. Species ratios for the domestic animals based on NISP (excluding dog) are shown in Table 3. The Razdolnoe assemblage is composed primarily of domestic cattle and horse bones; caprines make up only a small part of the assemblage; and pigs are very limited in number. The MNI estimates are also included in Table 3. These are minimum MNIs; the bones were not matched for size or age. The MNI data increase the relative importance of caprines. The sheep/goat assemblage yielded an MNI of 6 based on the mandibles. However, no caprine first, second, or third phalanges were recovered from the Razdolnoe faunal collection. No pig phalanges were recovered either. In contrast, the horse assemblage included 19 first phalanges, 17 second phalanges, and 11 third phalanges, and the cattle assemblage included 11 first phalanges, 8 second phalanges, and 4 third phalanges. While I do not have detailed information on excavation methods, the absence of smaller mammal phalanges and other smaller body parts such as carpals, tarsals, and patellae probably reflects the lack of fine sieving.

Table 1a. Animal species identified from the Late Bronze Age contexts at Razdolnoe.

Animal Species		N
Domestic mammals	Cattle (*Bos taurus*)	246
	Sheep (*Ovis aries*)	12
	Sheep/goat	53
	Pig (*Sus scrofa*)	3
	Horse (*Equus caballus*)	206
	Dog (*Canis familiaris*)	4
Wild mammals	Aurochs (*Bos primigenius*)	7
	Red deer (*Cervus elaphus*)	1
	Roe deer (*Capreolus capreolus*)	1
Small artiodactyl		1
Large artiodactyl		5
Large ungulate		81
Unidentified		832
Total		1452

Table 1b. Body-part distribution of cattle, sheep, sheep/goat, pig, and horse bones.

Anatomy	Cattle	Sheep	S/G	Pig	Horse
Skull	12	1	1		5
Horn Core	2				
Maxilla					1
Mandible	18	4	8		8
Hyoid	1				
Vertebrae	8		4		8
Rib	1				2

Continued

Anatomy	Cattle	Sheep	S/G	Pig	Horse
Os coxae	14		2		3
Femur	5		1	2	5
Patella	1				2
Tibia	8		3		4
Scapula	14		3		5
Humerus	3	2	3		3
Radius	10	1	7		4
Ulna	4				3
Astragalus	20	2	1	1	6
Calcaneus	8	1			8
Tarsals	5				8
Carpals	13				8
Metatarsals	10	1	1		8
Metacarpals	6		3		11
Metapodia	3				11
First Phalanx	11				19
Second Phalanx	8				17
Third Phalanx	4				11
Tooth fragments	3				3
Loose teeth	54		16		43
Total	246	12	53	3	206

Table 2. Measurements on domestic cattle and aurochs astragali from Razdolnoe compared with the range of measurements published for aurochs (Bökönyi 1995; see also Bartosiewicz et al 2006).

	GLl (mm)	Bd (mm)	N
Bökönyi Aurochs Ranges	77.0 – 97.0	51.0 – 69.0	
Razdolnoe Aurochs	83.6 – 87.2	52.1 – 60.2	2
Razdolnoe Domestic Cattle	60.4 – 73.2	38.4 – 48.6	8

Table 3. NISP, species ratios based on NISP, and MNI estimates for the Late Bronze Age domestic mammals from Razdolnoe.

	NISP	%NISP	MNI
Cattle	246	47.3	11
Sheep/goat	65	12.5	6
Pig	3	0.6	2
Horse	206	39.6	6

The faunal data from other sites in the Late Bronze Age of the Eastern European Steppe have been summarized by Morales Muñiz and Antipina (2009). The assemblages they surveyed are generally dominated by domestic cattle, with smaller numbers of horses and caprines, and very few pigs. Razdolnoe generally echoes this pattern, but the Razdolnoe faunal assemblage included a larger number of horse remains that most of the other contemporary assemblages. Like most of the assemblages surveyed by Morales Muñiz and Antipina, the remains of wild animals are quite rare at Razdolnoe.

The Razdolnoe assemblage is too small for detailed analysis of butchery practices or age profiles. However, the basic data presented here can contribute to our understanding of Late Bronze Age pastoral economies in eastern Ukraine. The Late Bronze Age economy was based on domestic animals, primarily cattle and horses, but this small sample also points to the use of wild forest-steppe forms, including deer and wild cattle. We plan to publish our complete dataset online to make it available to future researchers.

REFERENCES

Bartosiewicz, L., Boroneant, V., Bonsall, C. and Stallibrass, S. (2006) Size ranges of prehistoric cattle and pig at Schela Cladovei (Romania). *Analele Banatului* 14(1), 23–42.

Bökönyi S. (1995) Problems with using osteological materials of wild animals for comparisons in archaeozoology. *Anthropológiai Közlemények* 37, 3–11.

Campana, D. (2010) FAUNA: Database and Analysis Software for Faunal Analysis. Poster presented at the meeting of the International Council for Archaeozoology, August 23–26th, 2010, Paris.

Grant, A. (1982) The use of tooth wear as a guide to the ageing of the domestic ungulates. In: B. Wilson, C. Grigson and S. Payne (eds.) *Ageing and Sexing Animal Bones from Archaeological Sites*, 91–108. British Archaeological Reports, British Series 109. Oxford, BAR.

Kotova, N., Anthony, D., Brown, D., Degermendzhi, S. and Crabtree, P. (2017) The excavation of the Razdolnoe Site on the Kalmius River in 2010. In: S. Makhortykh and A. de Capita (eds.) *Archaeology and Palaeoecology of the Ukrainian Steppe*, 79–114. Kiev, Institute of Archaeology, National Academy of Sciences of Ukraine.

Morales Muñiz, A. and Antipina, E. (2003) Srubnaya faunas and beyond: A critical assessment of the archaeozoological information from the East European Steppe. In: M. Levine, C. Renfrew and K. Boyle (eds.) *Prehistoric Steppe Adaptation and the Horse*, 329–351. Cambridge, McDonald Institute for Archaeological Research.

Payne, S. (1973) Kill-off patterns in sheep and goats: The mandibles from Aşvan Kale. *Anatolian Studies* 23, 281–303.

Schmid, E. (1972) *Atlas of Animal Bones*. Amsterdam, Elsevier.

Silver, I. A. (1969) The ageing of the domestic mammals. In: D. Brothwell and E. S. Higgs (eds.) *Science in Archaeology*, 283–302. London, Thames and Hudson.

von den Driesch, A. (1976) *A Guide to the Measurement of Animal Bones from Archaeological Sites*. Peabody Museum Bulletin 1. Cambridge MA, Harvard University.

The Fauna from Lion Island, a Late Nineteenth and Early Twentieth Century Chinese Community in British Columbia, Canada

Shaw Badenhorst[1], Douglas Ross[2]

1. Evolutionary Studies Institute, University of the Witwatersrand, South Africa
2. Albion Environmental Inc., California, USA

Abstract: Lion Island is situated in the Lower Fraser River in British Columbia on the West Coast of Canada. Between 1885 and 1930, the Ewen Cannery operated from Lion Island. This was one of the largest salmon canning operations in British Columbia. During the summer season, the Ewen Cannery employed a labour force housed in spatially segregated camps. One of the camps located on Lion Island was where Chinese men lived in a two-story bunkhouse located adjacent to the canning complex. In 1901, a small community of Japanese fishermen and their families settled on an adjacent island, called Don Island, and supplied salmon to the cannery until its closure in 1930. After the closure of the cannery, most residents moved to the mainland. Like many other sites occupied by Chinese in western North America during the late nineteenth and early twentieth centuries, the faunal assemblage from Lion Island is dominated by pork. In addition, it contains a variety of species, and this aspect too is reflected in other sites occupied by Chinese in western North America at the time.

Keywords: Lower Fraser River, Canada, Lion Island, Chinese sites, pork

INTRODUCTION

A number of Chinese immigrants settled in various parts of the world during the 19th and 20th centuries. Many of these immigrants were employed as labourers, often living in racially distinct settlements. Fortunately, an increasing number of sites once inhabited by these communities are being excavated and studied archaeologically and the faunal remains are receiving more attention, including those of settlements located in western North America.

Gill (1985) studied fauna from Yema-po, a Chinese labour camp associated with construction of the Lake Chabot dam near San Leandro in California, beginning in 1874. Langenwalter and Langenwalter (1987: 45–51) summarised faunal studies on the late 19th century Riverside Chinatown in California. Longenecker and Stapp (1993) studied the fauna from Pierce, Idaho, a rural gold mining community inhabited between 1864 and 1932. Gust (1993) compared fauna from five Chinese sites: Sacramento (1850 – 1860), Woodland (1870 – 1880), Ventura (1890 – 1910) (all in California), Tucson (1880 – 1910) (Arizona) and Lovelock (1920 – 1930) (Nevada). Koskitalo (1995) studied the fauna from Barkerville, British Columbia dating from 1877 to 1947. These and other faunal studies are discussed and contextualised by Kennedy (2016), including his own recent work at the Market Street Chinatown in San Jose, California.

Previous archaeofaunal studies of overseas Chinese communities in western North America suggested that, typically, such sites are dominated by pork remains. In addition, find materials from overseas Chinese sites often consist of a relatively diverse range of species (Gill 1985; Gust 1993; Kennedy 2016; Koskitalo 1995; Langenwalter and Langenwalter 1987; Longenecker and Stapp 1993). However, it has been suggested that, in cases where Chinese communities were less well-off, less-expensive beef instead of pork as well as cheaper meat cuts dominate faunal samples (Gust 1993: 208).

THE EWEN CANNERY

Between 2005 and 2006, one of us (DR) excavated the remains of a small Chinese labouring community at an early industrial salmon cannery in British Columbia, Canada (Ross 2013) (Figure 1). Chinese migrants began arriving in substantial numbers in British Columbia in conjunction with the Fraser River gold rush of 1858, most originating from a small number of rural counties in the southeastern Chinese province of Guangdong (Li 1998). However, the largest influx occurred in the 1880s in response to the need for cheap labour in constructing the Canadian Pacific Railway. Besides mines and railroads, Chinese worked in a range of industries and established Chinatowns or Chinese quarters in settlements across the province, offering goods and services to Chinese and non-Chinese customers that included a range of imports from the homeland. The migrant population was largely male, although a small number of women were also present.

Industrial salmon canning in British Columbia developed in the early 1870s and was the primary focus of the West Coast fishing industry in the late nineteenth and early twentieth centuries (Newell 1988; 1991). Canneries were occupied on a seasonal basis between spring and fall, and typically comprised a series of wooden industrial and domestic structures raised on pilings over the intertidal foreshore of rivers and coastlines. Cannery labour was multiethnic, with housing spatially segregated along racial and ethnic lines. Chinese men and Japanese women comprised much of the labour force inside the cannery processing and canning fish while Japanese men worked primarily as fishermen, with European men serving as supervisors and business managers.

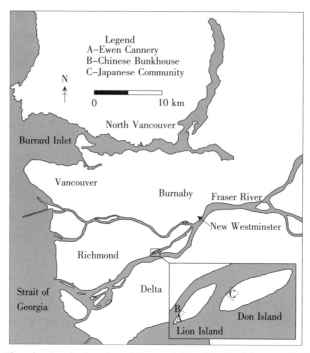

Figure 1. Plan showing the location of Don and Lion Islands (Drawing by D. Ross).

Between 1885 and 1930, the Ewen Cannery operated on Lion Island on the Lower Fraser River (Ross 2013). This was one of the largest industrial salmon canning operations in British Columbia. During the summer season, the Ewen Cannery employed a labour force housed in spatially segregated camps, organised along racial and ethnic lines. One of the camps located on Lion Island was where Chinese men lived in a two-story bunkhouse adjacent to the canning complex that could accommodate up to one hundred workers. The Chinese bunkhouse was built and owned by the cannery and seasonal labour was hired through Chinese labour contractors based in Vancouver's Chinatown, who provisioned the bunkhouse as part of the contract, often under exploitative circumstances. Unfortunately, little detailed information survives on the lives of bunkhouse residents at the Ewen Cannery, however, historical information on Chinese cannery labour in general suggests workers were at the low end of the economic spectrum. The constantly fluctuating Chinese labour force from season to season, a product of the contract system, did not foster development of an established domestic infrastructure, although it is likely residents kept a few pigs and chickens and tended a small garden. In 1901, a small community of Japanese fishermen and their families settled on an adjacent island, called Don Island, and supplied salmon to the cannery until its closure in 1930. After the closure of the cannery on Lion Island, most residents moved to the mainland.

The goal of the 2005 and 2006 fieldwork was to compare the daily lives of first-generation Chinese and Japanese immigrants to Canada with respect to patterns of cultural persistence and change and how they coped with life in a new and unfamiliar environment. Results indicate that they retained some elements of daily life from their homeland, but adopted other habits common in local Canadian society, albeit in unique ways (Ross 2013). The archaeofauna from Lion Island provides insights into the living conditions of Chinese male labourers during the late 19th and early 20th century in the Vancouver area on the West Coast of North America.

MATERIALS AND METHODS

Archaeological fieldwork conducted on Don and Lion Islands focused on identifying and excavating domestic midden deposits associated with the Chinese bunkhouse and Japanese settlement (Ross 2013). In the years since the cannery closed in 1930 all structures were removed from both islands, and the only substantial surface remains are of the main canning complex at the western tip of Lion Island, including rows of wooden pilings that once supported the principal factory buildings and adjacent wharfs (Plate XVI. Figure 1). Surface mapping and shovel testing at 5 metre intervals at the location of the former Chinese bunkhouse revealed two concentrations of domestic midden material. The East Midden was associated primarily with alcoholic beverage consumption and recreational activities and the West Midden with food preparation and consumption, including that of meat, although both categories of activities were represented in each midden (Figure 2). A series of 1 × 1 metre and 0.5 × 1 metre units was excavated in the two principal middens and elsewhere across the site to recover a sample of artifacts for analysis, using 10 centimetre levels because of a lack of identifiable stratigraphy

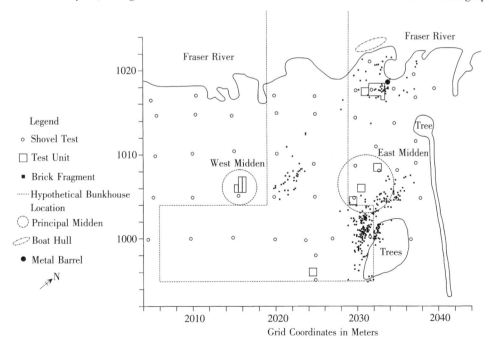

Figure 2. Plan of excavations at the Chinese bunkhouse on Lion Island (Drawing by D. Ross).

(Plate XVI. Figure 2). In total, 10.5 square metres of the site were excavated, with units ranging from 40 to 75 cm in depth depending on the thickness of cultural deposits. All excavated deposits were screened through a 1/4 inch (6 mm) wire mesh. Methodology developed by Driver (1992; 2005) was used to identify the faunal remains, albeit with a slight modification. All teeth were considered to be separate distinct elements and included in NISP counts (*pace* Driver 2005). The faunal remains were quantified using the Number of Identified Specimens (NISP) and Minimum Number of Individuals (MNI).

RESULTS

A total of 1218 bones and teeth were retrieved from the Chinese bunkhouse site on Lion Island (Plate XVI. Figure 3). A total of 356 specimens, representing 29% of the assemblage could be identified at some taxonomic level. The relatively high number of unidentified specimens (71%) is probably due to factors related to preservation, taphonomy and retrieval methods. The species composition is relatively diverse and includes pig (*Sus domesticus*), cattle (*Bos taurus*), indeterminate medium artiodactyls, medium mammal, large mammal, large artiodactyls, beaver (*Castor canadensis*), muskrat (*Ondatra zibethicus*), indeterminate small mammal, possibly rat (*cf. Rattus* sp.), indeterminate small rodent, indeterminate large, medium and small bird, a Galliformes, shark (Selachimorpha), possibly pikeminnow (*cf. Ptycocheilus*) and indeterminate fish (Table 1). The shark tooth is about 15 – 20 mm in length, and therefore from a large shark species. It is assumed that the indeterminate medium artiodactyla and medium mammals are probably pig, and that all indeterminate artiodactyla and large mammals are probably cattle.

Table 1. Lion Island faunal list (NISP and MNI).

Taxa	NISP	MNI
Species and Genus Identifications		
Sus domesticus (Pig)	145	5
Bos taurus (Cattle)	6	2
Ondatra zibethicus (Muskrat)	1	1
cf. *Rattus* (Possibly Rat)	4	1
Castor canadensis (Beaver)	2	1
cf. *Ptycocheilus* (Possibly Pikeminnow)	4	1
Class, Superorder and Order Identifications		
Artiodactyla Medium Indeterminate	8	
Artiodactyla Large Indeterminate	1	
Rodentia Small Indeterminate	1	
Mammalia Small Indeterminate	2	
Mammalia Medium Indeterminate	141	

		Continued
Taxa	NISP	MNI
Mammalia Large Indeterminate	6	
Galliformes	1	1
Aves Small Indeterminate	2	1
Aves Medium Indeterminate	17	2
Aves Large Indeterminate	1	1
Selachimorpha (Shark)	1	1
Pisces	13	1
Total NISP	356	18

A total of 206 bones were burnt, representing 17% of the entire sample. Six of these are burnt black, two have localised burning, three are burnt blue, two grey and 193 white. 131, or 11%, of the bones were either sawn or chopped through. Many of the saw marks are irregular in appearance with clear ridges caused by a metal saw. Another five have shallow chop marks, whilst 12 cut marks were noted. One large bird bone has a cut mark. Spiral fractures were present on 20 long bone fragments. One medium bird was represented by a spur, which are found on male birds, and common in Galliformes.

The proximal radius of an adult *Bos taurus* with a matching ulna fragment has peculiar damage. A hole, about 7 mm in diameter was drilled or struck between the medial humeral proximal articulation surface-glenoid cavity from/to the posterior portion of the proximal shaft of the radius. It barely missed the ulna. It may have been for a hook to suspend the carcass.

The sample contains one sub-adult cattle individual, 80 juvenile pigs, two juvenile indeterminate medium artiodactyls, 20 juvenile indeterminate medium mammals, one juvenile indeterminate small rodent and 32 juvenile unidentifiable bones. Based on post-cranial epiphyseal fusion data summarised by Bull and Payne (1982: 99), most, if not all, of the 80 juvenile pig remains are from individuals of less than one year of age. Many of the tooth fragments originate from deciduous teeth, un-erupted teeth as well as erupted teeth showing little or no wear. However, some adult pigs are also present. Only six pig specimens are clearly fused, and a single mandible with seven teeth has adult dentition. For cattle, only five specimens are clearly from adults, for indeterminate large mammal only two, and for indeterminate medium mammal only three specimens.

Skeletal representation was calculated for pigs (including indeterminate medium artiodactyls and indeterminate medium mammals). Most skeletal elements are represented (Table 2). Cattle, large artiodactyls and large mammal skeletal parts are listed in Table 3. Only a limited number of skeletal parts are represented from the faunal samples.

An adult left-sided cattle proximal radius is from a large individual. It was measured after von den Driesch (1976). The breadth of the proximal end (BP) measures 105.3 mm, and the proximal breath of the *facies articularis* (BFp) measures 96.9 mm.

Table 2. Lion Island skeletal elements for pig, medium artiodactyls and medium mammal (NISP).

Skeletal Element	Pig	Medium Artiodactyl	Medium Mammal	Total NISP
Cranium = 69				
Skull	1		4	5
Mandible	3		4	7
Maxilla	1		3	4
Teeth	51		2	53
Vertebrae, ribs, scapula and pelvic girdle = 129				
Atlas			1	1
Thoracic vertebrae			3	3
Lumbar vertebrae		1	16	17
Vertebrae indeterminate		5	18	23
Caudal vertebrae			1	1
Rib			72	72
Scapula	5		1	6
Innominate	2		4	6
Proximal limb bones = 28				
Humerus	16	1	5	22
Femur	4		2	6
Distal limb bones = 68				
Radius	5		2	7
Ulna	5		1	6
Tibia	6		1	7
Carpal	6			6
Tarsal	2			2
Astragalus	6			6
Metatarsal	1			1
Metapodial	8			8
Phalanx 1	4		1	5
Phalanx 2	13			13
Phalanx 3	3			3
Phalanx	3	1		4
Total	145	8	141	294

Table 3. Cattle, large artiodactyls and large mammal skeletal parts from Lion Island (NISP).

Skeletal element	Cattle	Large Artiodactyl	Large Mammal	Total
Thoracic vertebrae			1	1
Rib			5	5
Humerus	2	1		3
Radius	2			2
Ulna	2			2
Total	6	1	6	13

DISCUSSION AND CONCLUSION

Animal bones from 19th and 20th century Chinese sites in western North America display two distinct patterns: a predominance of pork remains, and a relatively diverse diet. This is hardly surprising, as in southern mainland China, where many immigrants originated, pork was at the time the most common meat source supplemented by a variety of other foodstuffs (see e.g. Service 1978: 441; Wang 1920: 289). First hand accounts indicate that this pattern often persisted among Chinese immigrants to North America in the late 19th and early 20th century (Brooks 1882; Peabody 1871: 661).

Gust (1993: 188) listed Canton prices from 1835 from most expensive to least expensive: mutton, hams (cured pork), goat, pork, sole fish, beef, pickled pig's feet, geese, chicken, duck, white rice, fish, oysters and eggs. In early 20th century China, Wang (1920: 290) indicates that fowl was more expensive than pork and fish. Even though pork was more expensive than beef (Longenecker and Stapp 1993: 110), Chinese immigrants in western North America continued to buy it. According to Wang (1920: 289), in China, the actual quantity of pork and other meats consumed are low. Meat is cut into small pieces and mixed with vegetables in a great variety of ways.

Faunal studies on Chinese sites in western North America found that pork dominates many faunal samples. This was noted at Yema-po (Gill 1985: 166), Riverside Chinatown (Langenwalter and Langenwalter 1987: 48), Barkerville (Koskitalo 1995: 18), Pierce (Longenecker and Stapp 1993: 118), and in Sacramento, Woodland, Ventura and Lovelock (Gust 1993: 180). However, at Tucson, beef bones outnumber pork by a large margin, ascribed to the inhabitant being less well-to-do (Gust 1993: 208). This pattern favouring beef is also reflected at other Chinese sites (Kennedy 2016: 136 – 142). In fact, based on a systematic review of faunal studies, Kennedy (2016: 77 – 86) concludes that meat consumption exhibits considerable heterogeneity within and between Chinese migrant communities, reflecting variables such as economics and class, local supply networks, relationships with non-Chinese individuals and communities, and emergence of new Chinese-American identities and consumer practices. Despite this complexity, pork was the most commonly consumed animal on Lion Island. The predominance of pork compared to less expensive beef at the Chinese bunkhouse is interesting, given that Chinese cannery workers were lower class laborers who made relatively little money. The likelihood that they raised pigs at the cannery, rather than purchasing pork from a butcher, may help explain this discrepancy.

The Lion Island faunal sample is dominated by juvenile domestic pigs of about one year of age based on tooth eruption, wear and post-cranial epiphyseal fusion. It was largely a common practice among Chinese butchers to sell pork of both juveniles and adults during the 19th and 20th centuries (*cf.* Langenwalter and Langenwalter 1987: 49; Longenecker and Stapp 1993: 115; Gust 1993: 191 – 193; Koskitalo 1995: 19). The predominance of juvenile pigs in the Lion Island faunal sample is therefore not unusual, and also makes sense if bunkhouse residents raised pigs on site.

19th and 20th century Chinese laborers often had access to more diverse foods than their Euro-American counterparts (Spier 1958a: 131). This is well-attested from archaeofaunal studies in western North America, including Lion Island. At sites such as Yema-po, Riverside Chinatown, Pierce, Sacramento, Woodland, Ventura, Tucson and Lovelock diverse species such as a wide range of ungulates, carnivores, rodents, lagomorphs, birds, reptiles, fish and molluscs are well-represented (Gill 1985: 165; Langenwalter and Langenwalter 1987; Gust 1993).

Typical products imported in the mid 19th century to North America by Chinese merchants include oranges, pumeloes, dry oysters, shrimps, cuttle fish, mushrooms, dry bean curd, bamboo shoots, narrow leaved greens, yams, ginger, sugar, rice, sweetmeats, sausage, dry duck, eggs, dried fruit, salt ginger, salt eggs and tea (Spier 1958b: 80). Such historical documents confirm the diversity of Chinese diets during the 19th and 20th centuries in western North America.

The beaver from Lion Island is not unexpected, as a breeding colony is still found on the island (DR personal observation, 2007). The probable house rat from Lion Island may have been eaten, although rats are attracted to human shelter for food. At Riverside, Langenwalter and Langenwalter (1987: 48) suggested that, based on breakage and cutmarks on rat bones, these animals may have been used as food or medicine. Gust (1993: 189) indicates that, in China, small vertebrates such as rats and mice were eaten by the poor.

Compared to other Chinese sites in western North America such as Yemo-po, Woodland Opera House and Riverside Chinatown (Gill 1985; Langenwalter and Langenwalter 1987: 47), very few bird bones were present in the Lion Island faunal sample. Gill (1985: 168) suggested that high numbers of chicken and other domestic fowl at these sites indicate the ability to keep domestic birds. This may suggest that at Lion Island, where birds were not intensely utilised, domestic birds were not kept at the site.

Eighteen fish bone fragments were recovered from Lion Island, which includes a shark tooth, and possibly pikeminnow. Spier (1958a: 81) mentions that seafood was a major contributor to South Chinese cookery during the 19th century. In California, species such as salmon, sturgeon, smelt, flounders, sculpins shrimp, abalone, crabs and oysters were used extensively (Spier 1958a: 83). Most seafood caught by Chinese was either salted or dried (Spier 1958b: 129; Brooks 1882). At Riverside for example, at least 28 different taxa are represented in the faunal sample (Langenwalter and Langenwalter 1987: 45), lending support to historical references to the importance of fish to Chinese immigrants.

For beef, the ranking of most expensive to least expensive meat cuts are: loin (pelvis ilium, thoracic and lumbar vertebrae), rib, round (femur), chuck (ribs, cervical, thoracic vertebrae, scapula, proximal humerus and humerus shaft), brisket (ribs), hindshank (tibia, tarsals), and foreshank (distal humerus, radius-ulna) (see Gust 1993: 185). Unfortunately, the beef (even including large artiodactyls and large mammals) sample from Lion Island is small, making inferences ambiguous. The bones from beef cuts at other Chinese sites in western North America suggests that these were bought from retailers (Gill 1985: 165; Longenecker and Stapp 1993: 110).

At Lion Island, less expensive chuck and foreshank beef cuts are found, with chuck cuts predominant. Whether this pattern is of any significance is unknown, as the small number of beef bones precludes any reliable interpretations. Moreover, it may not be reasonable to assume that Chinese labourers were never able to afford more expensive beef meat cuts over a period of many decades. The small number of beef remains may have been reserved for special occasions.

For pork, most to least expensive meat cuts are: loin (pelvis ilium, thoracic and lumbar vertebrae), ham (pelvis, femur, tibia), shoulder (scapula, humerus), picnic ham (scapula, humerus), belly (ribs) and jowl (head) (see Gust 1993: 185). Although some (Gust 1993; Koskitalo 1995) attempted to interpret pork meat cuts from Chinese sites in western North America as reflecting particular meat cut selection, the archaeofaunal analysis of Lion Island indicates that such attempts are unwarranted in light of various factors.

First, the 19th and early 20th century Chinese, both in mainland China and western North America, utilised all portions of pigs save for the hair and bones (Peabody 1871: 661; Wang 1920: 290). Previous faunal studies on Chinese sites in western North America noted that most if not all pork skeletal elements are represented in the samples (Langenwalter and Langenwalter 1987: 48; Longenecker and Stapp 1993: 115). This is also the case at Lion Island. Although some pork cuts were more expensive than others, it can be assumed that over various decades, most or all cuts were bought or obtained to suit local needs at different times. The almost complete representation of all pork elements suggests that, at times at least, complete carcasses were bought from retailers (Gill 1985: 165). However, the possibility that live pigs were kept on Lion Island cannot be excluded and is, in fact, probable given historical accounts of Chinese cannery workers raising pigs on site.

Second, bone cut numbers are inflated by the actual numbers of some elements in a single carcass, making comparisons ambiguous. For example, at Lion Island, ribs outnumber any other element by a large margin. However, most of the ribs are represented by small fragments only, therefore increasing their visibility and over-representing them in the sample. Langenwalter and Langenwalter (1987: 49) indicate that typical Chinese pork cuts include cleaver cut spare, short ribs of four to six inches (10 – 15 cm) and sectioned feet. Meat cuts may have been reduced even further so that they could be handled with chopsticks. However, Longenecker and Stapp (1993: 105 – 107) presented early documentary evidence that suggests that in San Francisco's Chinatown markets butchers paid no attention to different pork cuts, cutting flesh with cleavers indiscriminately.

Third, only 29% of the total Lion island sample was identified to element that was then assigned to taxa. Therefore, the unidentified bones inhibit estimating the true skeletal element frequencies of the sample (Badenhorst 2009; Badenhorst and Plug 2011; Lyman and O'Brien 1987). The high fragmentation levels at Lion Island are the result of various first and second order taphonomic processes that contributed to the fragmentary nature of the faunal sample (see summary in Reitz and Wing 1999).

According to Gust (1993: 206), little information is available on butchering as was practiced in China or the US during the middle to late 19th century. The large number of bones with sawn-and chop marks in the Lion Island sample, particularly among pig bones, indicates butchering and meat processing. Likely tools included knives, cleavers and hand and machine saws to butcher pig carcasses (Gill 1985: 165 – 166; Gust 1993: 193; Langenwalter and Langenwalter 1987: 48). Other equipment probably included hog hooks, bell-shaped scrapers and gambrels (Longenecker and Stapp 1993: 104 – 105). It is expected that similar tools produced the butchering evidence on bones in the Lion Island sample.

The Chinese male labourers from Lion Island chose to retain two distinctive features of their traditional diet that is evident from faunal remains: a predominance of pork, supplemented by a highly diverse diet. The continuation of traditional Chinese diet of pork meat was maintained despite higher retail prices for pork than beef and may have been accomplished in whole or in part by raising pigs on site.

However, recent research has begun to challenge assumptions about the significance of pork in the diet of Chinese immigrants. Kennedy (2016: 78) notes that earlier archaeological studies of Chinese migrant foodways "emphasize the continuation of supposedly traditional Chinese emphasis on pork consumption … and argue for the maintenance of Chinese food practices and thus a corresponding low degree of acculturation at these sites". He argues, however, that, while pork was common in regions of southern China affected by 19th century emigration, non-elite rural populations consumed it sparingly outside of feasts and festivals (Kennedy 2016: 60 – 61). Consequently, heavy consumption of pork by Chinese migrants in North America does not represent continuation of traditional southern Chinese dining habits but marks a distinct change in foodways. Such insights highlight the importance of moving "away from acculturative models of tradition versus change towards more nuanced understandings of Chinese migrant food practices and identities" (Kennedy 2016: 148).

ACKNOWLEDGEMENTS

We thank Dr. László Bartosiewicz from Stockholm University, Sweden for useful comments and suggestions.

REFERENCES

Badenhorst, S. (2009) Artiodactyla skeletal part representation at Middle Period and Early Plateau Pithouse Tradition sites on the Interior Plateau, British Columbia: A view from EdRh – 31. *Canadian Zooarchaeology* 26, 25 – 41.

Badenhorst, S. and Plug, I. (2011) Unidentified specimens in zooarchaeology. *Palaeontologia africana* 46, 89 – 92.

Brooks, W. (1882) A fragment of China. *Californian* 6(31), 6 – 15.

Bull, G. and Payne, S. (1982) Tooth eruption and epiphysial fusion in pigs and wild boar. In: B. Wilson, C. Grigson and S. Payne (eds.) *Ageing and Sexing Animal Bones from Archaeological Sites*. British Archaeological Series 109, 55 – 71. Oxford, BAR.

Driver, J. C. (2005) *Crow Canyon Archaeological Center Manual for Description of Vertebrate Remains*. Cortez, Crow Canyon Archaeological Center.

Driver, J. C. (1992) Identification, classification and zooarchaeology. *Circaea* 9(1), 35 – 47.

Gill, A. L. (1985) *A Pound of Pork and a Pinch of Puffer: Subsistence Strategies in a Chinese Work Camp*. MA thesis. Hayward, California State University.

Gust, S. M. (1993) Animal bones from historic urban Chinese sites: A comparison of Sacramento, Woodland, Tucson, Ventura, and Lovelock. In: P. Wegars (ed.) *Hidden Heritage. Historical Archaeology of the Overseas Chinese*. Baywood Monographs in Archaeology, 177 – 212. New York, Baywood Publishing.

Kennedy, J. R. (2016) *Fan and Tsai: Food, Identity, and Connections in the Market Street Chinatown*. Ph. D. dissertation. Bloomington, Indiana University.

Koskitalo, M. Y. (1995) *Faunal Analysis of a Historic Chinese site in Barkerville (FgRj – 1)*. BA Honours thesis. Burnaby, Simon Fraser University.

Langenwalter, P. E. and Langenwalter, R. E. (1987) An overview of the vertebrate fauna from Riverside's Chinatown. In: C. Brott (ed.) *Wong Ho Leun. An American Chinatown*. Vol. 2, 45 – 52. San Diego, The Great Basin Foundation.

Li, P. S. (1998) *Chinese in Canada*. Toronto, Oxford University Press.

Longenecker, J. G. and Stapp, D. C. (1993) The study of faunal remains from an overseas Chinese mining camp in northern Idaho. In: P. Wegars (ed.) *Hidden Heritage. Historical Archaeology of the Overseas Chinese*. Baywood Monographs in Archaeology, 97–122. New York, Baywood Publishing.

Lyman, R. L. and O'Brien, M. J. (1987) Plow-zone zooarchaeology: Fragmentation and identifiability. *Journal of Field Archaeology* 14(4), 493–498.

Newell, D. (1988) The rationality of mechanization in the Pacific salmon-canning industry before the Second World War. *Business History Review* 62, 626–655.

Newell, D. (1991) The industrial archaeology of the organization of work: A half century of women and racial minorities in British Columbia fish plants. *Material History Review* 33, 25–36.

Peabody, A. P. (1971) The Chinese in San Francisco. *The American Naturalist* 4(11), 660–664.

Reitz, E. and Wing, E. S. (1999) *Zooarchaeology*. Cambridge, Cambridge University Press.

Ross, D. (2013) *An archaeology of Asian transnationalism*. Gainesville, University Press of Florida.

Service, E. R. (1978) *Profiles in Ethnology*. London, Harper & Row.

Spier, R. F. G. (1958a) Food habits of nineteenth-century California Chinese (Part Two). *California Historical Society Quarterly* 37(2), 129–136.

Spier, R. F. G. (1958b) Food habits of nineteenth-century California Chinese (Part One). *California Historical Society Quarterly* 37(1), 79–83.

von den Driesch, A. (1976) *A Guide to the Measurement of Animal Bones from Archaeological Sites*. Peabody Museum Bulletin 1. Cambridge, Harvard University.

Wang, C. C. (1920) Is the Chinese diet adequate? *The Journal of Home Economics* 12(7), 289–293.

Archaeozoological and Historical Data on Sturgeon Fishing along the Danube

László Bartosiewicz[1], Gertrud Haidvogl[2], Clive Bonsall[3]

1. Osteoarchaeological Research Laboratory, Stockholm University, Sweden
2. Institute of Hydrobiology and Aquatic Ecosystem Management, University of Natural Resources and Life Sciences Vienna, Austria
3. School of History, Classics and Archaeology, University of Edinburgh, UK

Abstract: This article discusses archaeological and historical data on sturgeon (Acipenseridae family) in the Danube River with special emphasis on the great sturgeon (*Acipenser huso* Linnaeus, 1758 syn. *Huso huso* Brandt, 1869). Having established the complementary nature of information offered by prehistoric and medieval fishbone finds and the written record, it emphasizes that a multidisciplinary interpretive framework is indispensable in addressing ecological and economic questions involving traditional sturgeon exploitation, extinction and possible reintroduction in the Danube. Prior to their extinction in the Danube, sturgeons were affected by the sum of anthropogenic activities along the river's course, including increasing water transport, overfishing and the construction of dams. The detrimental effect of environmental damage can be seen in decreasing fish sizes and dwindling catch through time.

Keywords: Great sturgeon, Danube, archaeozoology, historical fishing, riverine environments

INTRODUCTION

The reconstruction of ancient habitats is an important step in the rehabilitation of present-day environments. Combining information obtained by two methodologically different directions of research, archaeological and historical inquiry has the potential to yield synergetic results when addressing human-environment interactions, which is a key element in researching environmental history. The *in situ* protection of endangered species and reintroduction of animals from hatcheries into areas they had previously inhabited would be enhanced by an in-depth understanding of their historical contexts. Great sturgeon (also known as beluga sturgeon) may be seen as a paradigmatic animal in this regard.

Within the framework of a broader project dedicated to studying interaction between the Danubian fish fauna and human societies, this paper focuses on multidisciplinary data on sturgeon fishing in various time periods at selected points along the river. Proceeding downstream, archival sources are presented for the Upper Danube region, concentrated in the Vienna area at ca. 1932 – 1924 river km from the Black Sea. In the Middle Danube Basin, both archaeological and documentary sources are analysed from Hungary, especially from the best researched northern section of the river stretching toward the Danube Bend (between ca. 1850 – 1698 river km) and near Budapest (1653 river km) where numerous excavations took place. Finally, prehistoric and Modern Age data are considered from the Iron Gates Gorge (1041 – 943 river km) between Romania and Serbia and its downstream exit section in the Lower Danube region.

There has been an increasing need to highlight methodological issues, especially the complementary nature of archaeoichthyological finds, environmental observation and the written historical record. Each of these is incomplete and represents not only different spatial and temporal foci, but also diverse approaches to the same problem: human impact on Acipenserid fish in the Danube with special emphasis on the beluga sturgeon (*Acipenser huso* Linnaeus, 1758). Various species in the Acipenseridae family were among the largest fish in the Middle and Lower Danube. Fishing regulations from the late Middle Ages and Early Modern times suggest that they frequently also occurred in the Upper Danube. Relatively recently, however, most of them have been brought to the brink of extinction by habitat loss, overfishing and the damming of the Iron Gates section in 1971 and 1984.

This latter development was especially detrimental to the species under discussion here, since with the exception of the smallest member of the Acipenseridae family, sterlet (*Acipenser ruthenus* Linnaeus, 1758), sturgeons are anadromous: during the spawning rush they swim upstream from the Black Sea into freshwater where they breed. Therefore, the construction of hydroelectric dams blocking the migration route has directly interfered with their reproduction cycle. Opening bypasses for sturgeon is only a partial solution as silting of the river bed (an inevitable consequence of decelerating current resulting from dam construction) also inhibits reproduction because a gravelly riverbed is needed as the fertilized eggs can stick to gravel but not silt. Sturgeon population decline is the result of a combination of factors, including inability to migrate and destruction of their habitat as well as overfishing and disturbances caused by riverine transport.

From a methodological point of view, it is interesting that migrating individuals may have been caught anywhere along this long route. The fish discussed in this paper thus represent the same population at different times, regardless of the location of archaeological sites where their bones were encountered.

CONSUMPTION: ARCHAEOZOOLOGICAL EVIDENCE

Most ordinary animal remains recovered from archaeological deposits represent food refuse. Therefore, the primary information gleaned from such finds concerns provisioning. In some cases, bones may have been left at butchery sites or removed from the area where the animal was eaten to be deposited elsewhere.

Prehistoric evidence discussed in this paper originates from the site of Schela Cladovei right on the left bank of the Danube, downstream from the Iron Gates gorge (present-day Romania) where these fish played a variegated role in the diet of the Late Mesolithic/Early Neolithic inhabitants of the region (Bonsall et al. 1997). It is possible that sturgeon fishing grounds were a major attraction in this area during prehistory. There have been speculations that some of the Mesolithic sculpted boulders recovered at Lepenski Vir (right bank, Serbia) upstream from Schela Cladovei actually depict sturgeons (Radovanovic 1997: 89).

Sturgeon bones recovered at these prehistoric settlements located directly on the riverside may reflect more-or-less localized consumption. It would be too speculative to assume that sturgeon meat was transported over long distances in the absence of tangible evidence at this early time.

The situation is different when we look at medieval bone finds of beluga sturgeon. Many of them have turned up far from the river, sometimes in hilly areas at high status settlements such as royal or ecclesial centres (Bartosiewicz and Bonsall 2008: 36, Figure 1). The relatively high frequency of medieval sturgeon bones at high-status sites, however, may also have gained special attention as medieval archaeology itself is more focussed on central settlements than river terraces where the best known prehistoric sites yielding sturgeon bones once existed.

Transport and distribution of the catch, salted or alive (Benda 2005: 253), is a cultural filter to be considered in the archaeological analysis of sturgeon remains. In the mid-18th century, Mátyás Bél described how large, live sturgeons were towed upstream by boat to markets in the capital cities of Buda, Hungary and Vienna, Austria (Bél 1764: 41).

Osteological evidence for perishable caviar, a particularly precious commodity mentioned in late medieval documentary sources (Kubinyi 2002: 249) does not exist, although chemically identifying proteins from fish roe in food residue has recently become possible (Shevchenko et al. 2018).

Complementing archaeozoological evidence qualitative information on direct consumption is available in written form in the few surviving Early Modern Age cookbooks where sturgeon recipes occur prominently in the list of species, sometimes referred to as "Royal Fish" (Rumpolt 1581; Galgóczi 1622).

PRODUCTION: FISHING SITES ALONG THE DANUBE

Directly related to the rehabilitation of present-day environments, the palaeohydrological reconstruction of riverine habitats in combination with some of the bone finds may pinpoint locations where sturgeon fishing could have been most successfully practiced. While animals weighing hundreds of kilograms could be harpooned or even caught with large metal hooks, some form of trapping made the use of any hand-held fish tackle easier. Nicolaus Olahus (1568) even mentions the complete closure of the Danube with such a structure, although a 1528 trial between the competing cities of Vác and Buda shows that this method was considered unacceptable. However, exploiting the natural hydrogeographic layout of rivers was a means readily available to all skilled fishermen whose specialized knowledge was probably handed down from generation to generation. Although sturgeon could be found throughout the Danube at different times of the year, diversity in the river's course created distinct opportunities for sturgeon fishing in some locations.

Horizontal variability: Until the mid-19th century, the Danube meandered through floodplains and wetlands, shifting its course with each spring and summer flood. Fish used such marshes, e.g. around Szentendre, north of Budapest, as resource rich hatcheries. In the plains, the variable course of the river resulted in underwater shoals, fords, and variously-sized islands that all influenced currents and created spots where sturgeons could be more easily caught. In 1690, "50 – 100 sturgeons were caught and butchered daily" at the island of Ada-Kaleh, downstream from Orşova (Romania) in the Iron Gates gorge (Marsigli 1726). This section of the Danube, including reaches downstream from the Iron Gates gorge, is shown in Marsigli's map reproduced in Figure 1.

Having plotted catch sites of 19th – 20th century record specimens on the map of present-day Hungary (Bartosiewicz and Bonsall 2008), this horizontal patterning appears to point to confluences and major river bends as good sturgeon fishing grounds. As rivers spread out downstream from major elevations in the riverbed, the web of shallow waters between sandbanks exposed large individuals moving in either direction.

Sturgeons were regularly caught in many of the Danube's tributaries, including the Váh, Maros and Tisza rivers (Hankó 1931: 9). In 1518, resulting from a long medieval tradition, the city of Komárom in northern Hungary was officially given the rank of Royal Sturgeon Fishing Grounds (Herman 1980: 267). This strategically important point is located at the confluence of two branches of the Danube, downstream from Europe's largest inland river delta where, according to osteological evidence, sturgeon had already been caught in Roman times (Bartosiewicz 1989: 611). The monopoly of Komárom fisherfolk can still be detected in the 18th century, when they leased sturgeon fishing rights from as far as the Tisza river between Tiszacsege and Tiszafüred (Bencsik 1970: 98), 250 km away in eastern Hungary.

Vertical variability: In addition to its course, the river's gradient also determined points where sturgeons could be easily targeted. When some historically renowned sturgeon fishing grounds are plotted along the entire section of the Danube (Figure 2), many of them pinpoint locations downstream from reaches with a steep gradient in the riverbed (large sturgeons were regularly caught at Tulln, upstream from Vienna, for 1605 and 1692 records are known from

Figure 1. The Danube downstream from the Iron Gates gorge in the 1726 book by Marsigli.
The major transversal line across the river marks the place where once Traian's bridge was built by Apollodorus of Damascus between AD 102 – 103. The prehistoric site of Schela Cladovei is located downstream from the northeastern bridgehead.

Bavaria; in 1616 a more than 100 kg specimen was caught in the Salzach River, a tributary of the Austrian Danube; Waidbacher and Haidvogl 1998). At the other end of the discussed section, the downstream exit of the Iron Gates gorge may also have served as a natural trap where upward migrating sturgeons, slowed by rapids, could be relatively easily caught already in prehistoric times.

Naturally, the distinction between horizontal and vertical elements is artificial; the best opportunities were offered by a three-dimensional combination of features. The fourth dimension, time, was also extremely important given the seasonal availability of sturgeons during the late spring and late fall, as two different migration regimes are known to exist for the great sturgeon in the Danube. One form migrates and spawns in the spring, while the other migrates during the fall, overwinters in the river, and moves further upstream to spawn during next spring. Chances of landing these fish also depended on the river's annual fluctuations related to weather conditions.

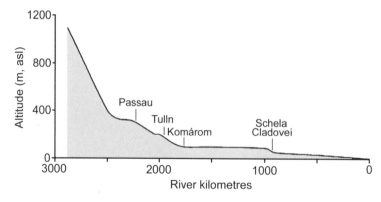

Figure 2. Vertical profile of the Danube with some locations of great sturgeon records.

Trade

In addition to consumption and production, documentary sources also concern trading and legal issues related to these majestic fish. The role of sturgeon was thus reviewed within the context of Early Modern Age and 18th to 20th-century changes of sturgeon populations in the Austrian Danube.

Sources from the Municipal and Provincial Archives of Vienna demonstrate that in the late 18th and early 19th centuries large quantities of sturgeons were sold in and around the city of Vienna. Trading of sturgeons was by then the privilege of the Viennese guild of fish traders, which separated from the guild of fishermen sometime between 1582 and 1620 (Matz and Tschulk 1979). According to a decree enacted by Emperor Karl VI in 1716, the civil fish traders were allowed to trade carp, pike, sturgeon, trout, catfish and eel. These fish came from the Danube and its tributaries in Lower and Upper Austria and Hungary as well as from Bohemia and Moravia. In terms of sturgeons, trading from Hungary can be assumed. Fish came to the market fresh, salted or smoked and were transported by ships or carriages. It is unlikely that sturgeons sold by civil fish traders were caught in the Viennese Danube because fishing there was the privilege of the Viennese guild of fishermen and local fishing communities around Vienna. The income from sturgeon and catfish trading was shared between all guild members. Thus, fish traders produced registers listing the weight of fish as well as their market value. The registers in the Viennese archives cover from 1795 until 1823 almost without gaps (Figure 3). They show fluctuating amounts of fish sold per year with a range from about 5 to the enormous amount of nearly 45 tons in the year 1796. Reasons for these fluctuations must await future research, but for the small amounts between 1805 and 1809, the Napoleonic Wars affecting trading along the Danube in Vienna and downstream could have played a role. Figure 3 also shows a decreasing trend in the total amount of traded fish within this period of about 30 years.

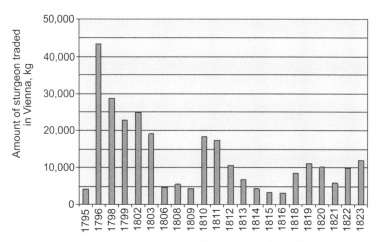

Figure 3. Sturgeon trade (total in kg) in Vienna during the 1795 – 1823 period.
Sources from the Viennese municipal archive: Wiener Stadt-und Landesarchiv, Bestand Fischkäufler 2. 8. 13 B13, Vol. 6.

The figures relating to Viennese sturgeon trading raise two questions: First, why trade sturgeon at all? Certainly, the demand for fish rose along with the number of the city's inhabitants. The Viennese population grew from approximately 30,000 in 1600 to 235,000 by the end of the 18th century. In addition, it is known from fish biological literature that sturgeons only rarely migrated as far as the Austrian Danube at the beginning of the 19th century (Fitzinger and Heckel 1835).

One can also assume that sturgeon stocks declined in the Austrian Danube already by the Early Modern Age due to intense fishing in the Upper as well as in the Middle Danube. So far, no historical sources have been identified enabling us to trace a decrease of sturgeon catches in the Austrian Danube directly from fishing records. Only modest information is available for the 18th century, e.g. in fish delivery records for the monastery of Klosterneuburg a few kilometers upstream from Vienna. Moreover, it is unclear if these fish were caught in the Austrian Danube or were imported from Hungary. But a decrease of sturgeon stocks in the Austrian Danube can be inferred from fishing regulations. For the city of Tulln there is information about sturgeon fisheries dating back to the 10th century AD (Raab 1978). Mentions of sturgeons in local bylaws, e.g. for the Danube villages of Höflein, Klosterneuburg and Vienna (communities of Erdberg and Scheffstraße), from the 14th to the 16th centuries indicate that catches were regularly anticipated. According to these documents fishermen were obliged to deliver sturgeons to the monastery of Klosterneuburg or to the imperial court, which is also evidence for the high status of these fish (Winter 1886). From the terms "*Hausen*" and "*Dyck*" it can be assumed that in the Austrian Danube either the beluga sturgeon or the Russian sturgeon (*Acipenser gueldenstaedtii* Brandt & Ratzeburg, 1833) was most commonly caught. Local regulations appear until the 16th century. In contrast to recurring mentions of sturgeons in local regulations, they are never considered in country-wide fishing laws enacted for the Austrian section of the Danube in the 16th century and thereafter (Hoffmann and Sonnlechner 2011).

A second indirect hint of decreasing sturgeon stocks in the Austrian Danube was the intensification of sturgeon fishing in the Middle Danube and the subsequent decrease of sturgeon stocks in the 16th century (Balon 1968). Sturgeons from the Middle Danube were traded not only to Vienna but also to Prague, Munich, Krakow and even Paris. As a consequence, Balon (1968) noted a decrease in sturgeon catches in the 17th and especially in the 18th century.

Sturgeon sizes

A prominent feature of beluga sturgeons is their large size. Antipa (1905) refers to an extreme record in 1890 when a beluga sturgeon of 882 kg was caught in the mouth of the south branch of the Danube Delta known as Sfäntu Gheorghe.

Figure 4. Sturgeon prior to butchery. Lower Danube region. Proportions indicate that even if the fisherman was only 165 cm tall, this specimen would have been 2.7 m long (Source: http://www.slideshare.net/acvabio/dunarea-veche-6052418).

Despite a gradual but consistent diachronic size decrease, even in recent centuries some of these animals grew to spectacular sizes. Records exist of specimens measuring over 3 metres, such as the great sturgeon shown in Figure 4 most probably photographed prior to the damming of the Iron Gates gorge sometime during the third quarter of the 20th century.

Information on size can be obtained from both archaeological finds and written documents. This trait, therefore, represents a common denominator between the two types of sources analyzed in this study. The only difference is that while bone remains may represent broad size ranges, sizes reported in writing almost invariably concern the largest specimens.

The first, most robust, pectoral fin ray was recommended for use in the osteometric reconstruction of common sturgeon (*Acipenser sturio* Linnaeus, 1758; Desse-Berset 1994: 84). The greatest length of beluga sturgeon could be likewise estimated from the mediolateral width of this bone (Bartosiewicz and Takács 1997: 12, Figure 8/1). Length estimates for prehistoric specimens from Schela Cladovei and medieval sturgeons from Hungary (Bartosiewicz and Bonsall 2008) were plotted together with Modern Age historical data (Khin 1957) in Figure 5.

The distribution of pooled archaeological length estimates and Modern Age records displays a slight positive skew but largely within the same range, suggesting that prehistoric sturgeons

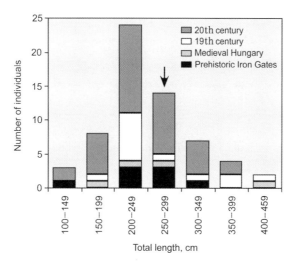

Figure 5. The distribution of total lengths of sturgeon caught in the Danube.
The graph combines estimates made on the basis of prehistoric and medieval bone finds with 19th – 20th century written records. The arrow indicates the size interval relevant to the specimen shown in Figure 4, not included in the calculations.

reconstructed from the Schela Cladovei finds were as large as the largest modern specimens in the recent past of Hungary. In order to test the methodological difference between these two primary datasets, the mean value of total lengths estimated from archaeological finds was compared to the mean value of Modern Age written records in Table 1 using a Student's t-test.

Table 1. Comparing archaeological estimates and historical records of total lengths (cm) of sturgeon using a Student's t-test

	Archaeological estimates	Modern record specimens
Number of specimens	14	47
Mean	236	257
Standard Deviation	83.6	62.9
Maximum	425	425
Minimum	125	175
Median	225	225
t-value		−1.02
p-value		0.31
Degrees of freedom		59

The distribution of length estimates for the largest prehistoric sturgeons corresponds to those of the longest specimens from historical periods. The span's worth of difference in mean values is not significant on the required $p \leq 0.05$ level of probability. The only difference between the two samples is that the small set of archaeological specimens includes smaller individuals as well. This is not only shown by the smaller minimum value of archaeological fish but also by the smaller standard deviation of the modern historical sample focused on record specimens.

However, the overwhelming majority of sturgeons represented by osteological measurements seem to have been as long as Modern Age record individuals. This similarity is probably also an artefact of differential preservation in the soil. The effect of selective recovery of large bones may be ruled out as the Schela Cladovei material was recovered with water-sieving using a 1 mm mesh.

Previous analyses, however, showed that the virtual extinction of beluga sturgeon in the Danube was preceded by at least 150 years of decline in the bodyweight of record specimens: an average annual decrease of 1. 84 kg in record sturgeon body weights since 1850 was found to be statistically significant (Bartosiewicz and Takács 1997: 9). This reduction is mirrored by a similarly pronounced decline in the overall weight of the sturgeon catch in Romania during the 20th century (Bartosiewicz et al. 2008: 41, Figure 3). Figure 6 illustrates this clear tendency. The graph also shows the position of an unexpectedly large 181 kg specimen caught at Paks, Hungary in 1987, nearly two decades after the closure of the Iron Gates 1 Dam (Pintér 1989: 24). This animal may have attained its unusually large size after being trapped upstream behind the dam.

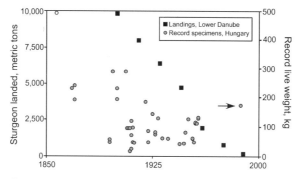

Figure 6. The parallel decline in record sizes in Hungary and total sturgeon landings in Romania preceding the damming of the Iron Gates section.
The large 1987 specimen caught at Paks is indicated by an arrow.

Perspectives in sturgeon rehabilitation

Large-bodied, slowly reproducing acipenserid fish have been threatened by extinction across Eurasia (e.g. Pegasov 2009; Du et al. 2013; Friedrich 2018). Overfishing, incessant habitat degradation, and the concomitant decline of genetic integrity in fragmented populations has taken its toll. Today great sturgeons can swim into the 943 km long Lower Danube, up to the Iron Gates. Valuable propagation and stocking programs are carried out in this section of the river. Upstream from this point, however, sturgeons are extinct.

Sturgeon conservation is a complex long-term task that can only be carried out on a multinational level. In order to promote sturgeon rehabilitation the *Action plan for conservation of sturgeons* (*Acipenseridae*) *in the Danube River Basin* (Bloesch et al. 2005) was launched within the framework of the "Bern Convention" in 2005 and adopted in 2018. A ten years long total ban on sturgeon fishing was implemented by 2005 in Romania and Bulgaria soon followed suit. However, poaching and illegal fishery still threaten the carefully cultivated stocks. In 2015 the ban was extended by another five years, as completing a viable rehabilitation program may take at least two sturgeon generations, no less than 25 – 50 years. Unfortunately, financing decisions often need to be made in cycles shorter by an entire order of magnitude (2 – 5 years).

In addition to the costly *in situ* propagation of sturgeons, further efforts would be required to facilitate migration by allowing the anadromous fish to overcome artificial barriers in the river. Helping sturgeons to get around the two Iron Gates dams could reopen over 1000 river km of habitat (Friedrich 2018) by providing access to the Middle Danube within the Carpathian basin. Although this construction work would require massive industrial investment, archaeological and historical evidence from three countries suggests that successfully re-stocking sturgeon populations might benefit not only conservation but, on a longer run, even local economies.

CONCLUSIONS

Having established the complementary nature of information offered by archaeological fishbone finds and the written record, this study emphasizes that a multidisciplinary interpretive framework is indispensable in addressing ecological and economic questions involving traditional sturgeon exploitation, extinction and possible reintroduction in the Danube.

This is true especially for diadromous species that are affected by the sum of anthropogenic activities along a river's course. These included intensifying riverine traffic, the construction of dams changing aquatic habitats and blocking migration routes as well as a traditional focus on highly valued, large sturgeons that may have resulted in early overexploitation caused by intensive fishing. These historical lessons should be taken into consideration in the future rehabilitation of sturgeon populations.

REFERENCES

Antipa, G. (1905) Die Störe und ihre Wanderungen in den europäischen Gewässern mit besonderer Berücksichtigung der Störe der Donau und des Schwarzen Meeres. *Sitzungsberichte des internationalen Fischerei-Kongresses in Wien*, 1 – 22.

Balon, E. (1968) Einfluß des Fischfangs auf die Fischgemeinschaften der Donau. *Archiv für Hydrobiologie*, Suppl. 34(3), 228 – 249.

Bartosiewicz, L. (1989) Animal remains from the Fort. In: D. Gabler (ed.) *The Roman Fort of Ács-Vaspuszta* (*Hungary*) *on the Danubian Limes. Part II*, 600 – 623. Oxford, British Archaeological Reports.

Bartosiewicz, L. and Takács, I. (1997) Osteomorphological studies on the Great Sturgeon (Huso huso Brandt). *Archaeofauna* 6, 9 – 16.

Bartosiewicz, L., Bonsall, C. and Șișu, V. (2008) Sturgeon fishing along the Middle and Lower Danube. In: C. Bonsall, V. Boroneanț and I. Radovanović (eds.) *The Iron Gates in Prehistory. New perspectives*, 39 – 54. Oxford, Archaeopress.

Bartosiewicz, L. and Bonsall, C. (2008) Complementary taphonomies: Medieval sturgeons from Hungary. In: Ph. Béarez, S. Grouard and B. Clavel (eds.) *Archéologie du Poisson. 30 ans d'archéo-ichtyologie au CNRS. Hommage aux travaux de Jean Desse et Nathalie Desse-Berset*. Antibes, XXVIIIe rencontres internationales d'archéologie et d'histoire d'Antibes, 35 – 45. Éditions APDCA.

Bél, M. (1764) *Tractatus de rustica Hungarorum: A magyarországi halakról és azok halászatáról* [Fish and Fishing in Hungary]. Vízügyi Történeti Füzetek 73. Reprinted in 1984, Budapest.

Bencsik, J. (1970) Egy jobbágyközség gazdasági, társadalmi élete az úrbérrendezéstöl a jobbágyfelszabadításig [The economic and social life of a serfs' village from property regulations until liberation]. *Acta Universitatis de Ludovico Kossuth Nominatae Series Historica* 10, 49 – 91.

Benda, B. (2005) *Étkezési szokások a 17. századi főúri udvarokban Magyarországon* [Eating habits in 17th century Hungarian aristocratic courts]. PhD thesis, Budapest, Eötvös Loránd University. http://archivum.piar.hu/batthyany/benda/doktori-disszertacio.pdf

Bloesch, J., Jones, T., Reinartz, R. and Striebel, B. (2005) Action plan for the conservation of the sturgeons (Acipenseridae) in the Danube River basin. Strasbourg, Convention on the Conservation of European Wildlife and Natural Habitats.

Bonsall, C., Lennon, R., McSweeney, K., Stewart, C., Harkness, D., Boroneanț, V., Bartosiewicz, L., Payton, R. and Chapman, J. (1997) Mesolithic and Early Neolithic in the Iron Gates: A palaeodietary perspective. *Journal of European Archaeology* 5(1), 50 – 92.

Desse-Berset, N. (1994) Sturgeons of the Rhône in Arles (6th – 2nd century BC). In: W. Van Neer (ed.) *Fish Exploitation in the Past*, 81 – 90. Tervuren, Koninklijk Museum voor Midden-Afrika, Annalen, Zoologische Wetenschappen Vol. 274.

Du, H., Wang, C., Wei, Q., Zhang, H., Wu, J. and Li, L. (2013) Distribution and movement of juvenile and sub-adult Chinese sturgeon (*Acipenser sinensis* Gray, 1835) in the Three Gorges Reservoir and the adjacent upstream free-flowing Yangtze River section: A re-introduction trial. *Journal of Applied Ichthyology* 29, 1383 – 1388.

Fitzinger, L. and Heckel, J. (1835) Monographische Darstellung der Gattung Acipenser. *Annalen des Wiener Museums der Naturgeschichte* 1, 261 – 326.

Friedrich, T. (2018) Danube Sturgeons Past and future. In: S. Schmutz, and J. Sendzimir (eds.) *Riverine Ecosystem Management*. Aquatic Ecology Series, Vol. 8, 507 – 518. Cham, Springer.

Galgóczi, I. (1622) *Szakácsi Tudoman* [Cooks' Science]. Manuscript.

Hankó, B. (1931) Magyarország halainak eredete és elterjedése [The origins and distribution of fishes in Hungary]. *A Debreceni Tisza István Tudományegyetem Állattani Intézetéből* 10, 3 – 34.

Herman, O. (1980) *Halászat és pásztorélet* [Fishing and pastoral life]. Budapest, Gondolat Kiadó.

Hoffmann, R. and Sonnlechner, Ch. (2011) Vom Archivobjekt zum Umweltschutz. *Jahrbuch des Vereins für Geschichte der Stadt Wien* Bd. 62/63, 79 – 133.

Khin, A. (1957) *A magyar vizák története* [The history of sturgeons in Hungary]. Budapest, Mezőgazdasági

Múzeum Füzetei 2.

Kubinyi, A. (2002) Főúri étrend tábori körülmények között 1521 – ben [Aristocratic diets in military camps in 1521]. In: P. Fodor, G. Pálffy and I. Gy. Tóth (eds.) *Tanulmányok Szakály Ferenc emlékére*, 249 – 261. Budapest, MTA TKI Gazdasági-és Társadalomtörténeti Kutatócsoport.

Marsigli, L. F. (1726) *Danubius Pannonico-Mysticus*. Vol. IV. Amstelodam, Uytwerf & Changuion.

Matz, H. and Tschulk, H. (1979) Fischerei im alten Wien. *Wiener Geschichtsblätter*, Beiheft 4, 1 – 8.

Olahus, N. (1568) Hungaria. In: I. Sambucus (ed.) *Antonio Bonfini, Rerum Ungaricarum decades*. Basilea.

Pegasov, V. A. (2009) The ecological problems of introduction and reintroduction of sturgeons. In: R. Carmona, A. Domezain, M. García Gallego, J. A. Hernando, F. Rodríguez and M. Ruiz-Rejón (eds.) *Biology, Conservation and Sustainable Development of Sturgeons*, 339 – 343. Cham, Springer.

Pintér, K. (1989) *Magyarország halai* [The Fish of Hungary]. Budapest, Akadémiai Kiadó.

Raab, A. (1978) *Die traditionelle Fischerei in Niederösterreich mit besonderer Berücksichtigung der Ybbs, Erlauf, Pielach und Traisen*. PhD thesis, Wien, Universität Wien.

Radovanović, I. (1997) The Lepenski Vir Culture: A contribution to the interpretation of its ideological aspects. In: M. Lazić (ed.) *Antidoron Dragoslavo Srejovic completis LXV annis ad amicis, collegis, discipulis oblatum*, 87 – 93. Belgrade, Center for Archaeological Research.

Rumpolt, M. (1581) *Ein new Kochbuch*. Franckfort am Mayn, Verlag Sigmund Feyerabendt.

Shevchenko, A., Schuhmann, A., Thomas, H. and Wetzel, G. (2018) Fine Endmesolithic fish caviar meal discovered by proteomics in foodcrusts from archaeological site Friesack 4 (Brandenburg, Germany). *PloS one* 13(11), e0206483.

Waidbacher, H. and Haidvogl, G. (1998) Fish migration and fish passage facilities in the Danube: Past and present. In: M. Jungwirth, S. Schmutz and S. Weiss (eds.) *Fish Migration and Fish Bypasses*, 85 – 98. Oxford, Fishing News Books.

Winter, G. (ed.) (1886) *Niederösterreichische Weisthümer I., Das Viertel unter dem Wienerwald*. Wien, Österreichische Weistümer Vol. 7.

吉林大安后套木嘎遗址出土贝类遗存研究

The Study of Shell Remains from the Houtaomuga Site in Da'an, Jilin

陈全家　　王雅艺　　王春雪
Quanjia Chen　Yayi Wang　Chunxue Wang

吉林大学边疆考古研究中心，长春
Research Center for Chinese Frontier Archaeology of Jilin University, Changchun, China

摘要：本文以2011~2015年后套木嘎遗址出土贝类遗存为研究对象，进行动物考古学研究。通过对该遗址贝类遗存的时空分布规律及出土贝制品进行系统研究，提取出了大量的人类行为信息，为复原该遗址古代居民生产生活的整体面貌提供了一个独特的视角。

关键词：贝类遗存；后套木嘎；时空分布；贝制品

Abstract: The site of Houtaomuga (HTMG) is located in Da'an City, Jilin Province, China. It was excavated between 2011–2015, by archaeologists at the Research Center for Chinese Frontier Archaeology of Jilin University and the Jilin Provincial Institute of Archaeology. The site history can be divided into seven periods, from the Neolithic to the Liao and Jin dynasties. A large number of shells were unearthed during excavations at the site, providing multiple clues to ancient human behaviors. We studied the shells gathered from the five-year excavation of HTMG using zooarchaeological research methods. The results are presented in this article.

The article is organized as follows:

Section 1 describes the general situation of the shell remains at the site.

In Section 2, the temporal and spatial distribution of shells at the site is investigated. Statistical analysis was carried out to examine time and space distributions to reveal patterns in the way shells appeared on the settlement and how they were used in these seven period represented by various types of archaeological features (strata, remains of houses, pits, ditches, and burials). The shells at the site are divided into two main categories: The Anodontinae species were mostly used to manufacture shell artifacts; the other taxa provide information about the proportion of shell species brought back to the site. The two groups of

---

* 本研究由2015年"国家社科基金重大项目"（15ZDB55）资助。

specimens were counted individually.

Section 3 goes more deeply into the shell artifacts that came to light from the site excavations. A large number of semi-finished shell artifacts made of Anodontinae were analyzed typologically along the operational chain. An analysis of the typology, production, and function of the shell artifacts unearthed from the site were carried out.

Conclusions appear in Section 4.

Based on the analysis of the temporal and spatial distribution of shells at HTMG, the following three conclusions were drawn. Firstly, over time, the shell artifacts production left less and less waste and the manufacturing technology became more and more efficient. Secondly, the shell species and their proportions at the site did not display significant changes across different periods, suggesting a stable natural environment. Thirdly, the proportion of mollusk taxa from different types of features varies, indicating that the people at the settlement consciously chose different mollusk taxa according to their needs.

The shell artifacts at the site show the following characteristics: Firstly, different taxa of shell artifacts display different manufacturing profiles. Secondly, the inhabitants produced shell artifacts, adapting to the morphological features of different kinds of shells, making full use of each portion of the shell. Thirdly, the utilization of shells at the site was complex. The same type of shell material could be used as a tool or refined into a delicate ornament.

From the quality of production it seems that the inhabitants of the settlement had mastered manufacturing techniques including direct percussion, abrasion, cutting, scraping and piercing. They applied these techniques to both the manufacture of shell tools and ornaments. The tool-making ability of the population was relatively mature while a large number of their shell products had become standardized. They used different shells to make different objects. They also had a clear sense of the shape and size of each type of object, with a developed aesthetic resulting in the production of exquisite ornaments.

Based on the study of the shell remains from Houtaomuga, a preliminary picture has emerged of the way mollusk resources in this region were utilized over the seven periods when the settlement was in use. Close examination of mollusks sheds light on reconstruction of the past lifeways of the people who lived at this ancient settlement.

Keywords: shell remains, Houtaomuga, distribution in space and time, shell artifacts

后套木嘎遗址位于吉林省大安市红岗子乡永合村西北的漫岗中段，新荒泡的东南岸。遗址海拔152米，岗顶高出西北侧湖面约6~12米。遗址面积141万平方米，遗存密集分布区55万平方米[1]。

2011~2015年，吉林大学边疆考古研究中心和吉林省文物考古研究所对后套木嘎遗址进行了大规模联合发掘，五年的考古工作中揭露面积6450平方米。该遗址出土遗存大致可以分为七期：第一期："后套木嘎第一期类型"遗存，属于新石器时代早期

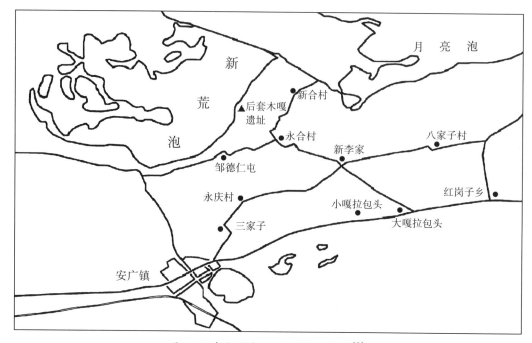

图 1 后套木嘎遗址地理位置示意图[2]
Figure 1. Location of the Houtaomuga site.

(13010~8650 BC)；第二期："黄家围子类型"遗存，年代为距今 8000~7000 年前后；第三期："后套木嘎第三期类型"遗存，年代为距今 6500~5600 年前后；第四期：类似于哈民忙哈文化遗存，年代为距今 5500~5000 年，相当于红山文化晚期；第五期：类似于白金宝文化晚期遗存，年代大致为西周晚期至春秋时期（877~476 BC）；第六期：类似于汉书二期文化遗存，年代为战国至西汉时期（475 BC~AD 8）；第七期：年代为辽金时期（AD 907~1234）[3]。

后套木嘎遗址出土了大量软体动物壳标本，经过筛选、整理，发现其显示出大量人类行为信息。本文运用动物考古学研究方法对后套木嘎遗址五年发掘所得的软体动物壳标本及贝制品进行系统研究。

一 贝类遗存出土概况

后套木嘎遗址出有七种软体动物，双壳纲包括无齿蚌亚科（Anodontinae，图版 17：1）、圆顶珠蚌（*Unio douglasiae*，图版 17：3）、短褶矛蚌（*Lanceolaria glayana*，图版 17：5）、河蚬（*Corbicula fluminea*，图版 17：4）及虾夷盘扇贝（*Mizuhopecten yessoensis*，图版 17：2）五种，腹足纲包括中国圆田螺（*Cipangopaludina chinensis*，图版 17：6）、纵肋织纹螺（*Nassarius variciferus*，图版 17：7）、灰巴蜗牛（*Bradyaena ravida*，图版 17：8）三种，其中虾夷盘扇贝和纵肋织纹螺为海水生贝类。

遗址中贝壳标本出土数量巨大，但由于自然和人为因素影响，标本大多残破不堪，尤以无齿蚌亚科为甚。

出土无齿蚌亚科标本逾万件，受风化作用等影响，整体保存情况差，不见完整标本，均为残片，无法对其进行部位及最小个体数的统计，仅选取2217件人工制品进行研究。其中2140件为蚌料，仅初步加工，未经定型，难辨器形；77件为蚌器及残件，经精细打制、磨制定型，部分见有穿孔痕迹，可辨器形。

除无齿蚌亚科外的其他贝类保存情况较好，多可鉴定方位及部位，共6861件（表1）。其中圆顶珠蚌数量最多，最小个体数占总数的86.86%；其次是短褶矛蚌，占8.74%；中国圆田螺占3.81%；其余种属仅占0.59%。

表1 其他贝类标本出土概况（单位：件）
Table 1. NISPs and MNIs of mollusk taxa (except Anodontinae) from HTMG.

	种属 Taxa		可鉴定标本数 NISP		最小个体数 MNI
双壳纲	圆顶珠蚌 *Unio douglasiae*	左 L	3050		3050
		右 R	3042		
	短褶矛蚌 *Lanceolaria glayana*	左 L	309		309
		右 R	304		
	河蚬 *Corbicula fluminea*	左 L	1		3
		右 R	2		
	虾夷盘扇贝 *Mizuhopecten yessoensis*	左 L	1		1
		右 R	0		
腹足纲	中国圆田螺 *Cipangopaludina chinensis*		135		135
	纵肋织纹螺 *Nassarius variciferus*		13		13
	灰巴蜗牛 *Bradyaena ravida*		4		4
	合计 Sum		6861		3515

贝制品共2260件，其中无齿蚌亚科制品2217件，短褶矛蚌制品34件，圆顶珠蚌制品9件，虾夷盘扇贝制品1件。种类有蚌刀、蚌匙、刻划器、片状蚌饰、穿孔蚌饰、蚌珠串饰、三棱形蚌饰、矛形蚌饰等。

二 贝类遗存时空分布

为充分了解该遗址贝类遗存，从时间、空间分布对其进行各项统计分析，以探知遗址不同分期（第一期至第七期）、各种单位类型（地层、房址、灰坑、灰沟及墓葬）中贝类遗存分布及其所含信息。

据出土情况将遗址中的贝类分为两类：其一，是主要体现加工行为的无齿蚌亚科；

其二，是主要体现种属比例信息的其他贝类。为更明确、清晰地提取贝类遗存包含的信息，对这两类标本分别进行统计。

（一）时间分布

该遗址分期明确并出有贝类遗存的单位共计 275 个（表 2），这些单位中共出土贝类标本 5744 件，包括无齿蚌亚科标本 1257 件，其他贝类标本 4487 件。因第五期遗存仅有一个单位，且出土样本量过少，为避免对分析结果产生误导，不纳入统计范畴。

表 2　分期明确并出有贝类遗存的单位数量（单位：个）
Table 2. Numbers of features with clear period and presence of shell remains.

单位 分期	房址 House	灰坑 Pit	灰沟 Ditch	墓葬 Burial	合计 Sum
第一期 Period Ⅰ	0	8	4	1	13
第二期 Period Ⅱ	1	32	2	6	41
第三期 Period Ⅲ	4	83	3	5	95
第四期 Period Ⅳ	22	61	6	1	90
第五期 Period Ⅴ	1	0	0	0	1
第六期 Period Ⅵ	1	0	0	27	28
第七期 Period Ⅶ	0	4	3	0	7
合计 Sum	29	188	18	40	275

1. 无齿蚌亚科

该统计涉及 1257 件标本，其中 1215 件为蚌料、32 件为蚌器及残件。对蚌制品成品率进行计算，结果如表 3 所示。

表 3　无齿蚌亚科标本各期成品率（单位：件）
Table 3. Rates of finished product of Anodontinae per period.

	第一期 Period Ⅰ	第二期 Period Ⅱ	第三期 Period Ⅲ	第四期 Period Ⅳ	第六期 Period Ⅵ	第七期 Period Ⅶ
蚌料 Semi-finished product	485	88	209	413	12	8
蚌器成品及残件 Finished product	10	1	10	19	1	1
合计 Sum	495	89	219	432	13	9
成品率 Rate of finished products	2.02%	1.12%	4.57%	4.40%	7.69%	11.11%

由表 3 可见，无齿蚌亚科制品成品率从早到晚总体呈上升趋势。随时间发展，加工蚌器产生的废料越来越少，加工工艺日趋成熟。

2. 其他贝类

对 4467 件其他贝类标本进行分期统计，结果如表 4 所示。

表 4 其他贝类各期出土情况（单位：件）

Table 4. NISPs of mollusk taxa (except Anodontinae) from HTMG (with clear period).

		第一期 Period I	第二期 Period II	第三期 Period III	第四期 Period IV	第六期 Period VI	第七期 Period VII
圆顶珠蚌 *Unio douglasiae*	左 L	112	178	631	1049	19	60
	右 R	92	164	684	1050	21	62
短褶矛蚌 *Lanceolaria glayana*	左 L	1	29	71	72	8	2
	右 R	2	15	56	64	11	2
河蚬 *Corbicula fluminea*	左 L	0	0	1	0	0	0
	右 R	0	0	1	1	0	0
虾夷盘扇贝 *Mizuhopecten yessoensis*		0	0	0	0	0	0
中国圆田螺 *Cipangopaludina chinensis*		2	3	5	10	4	0
纵肋织纹螺 *Nassarius variciferus*		1	0	1	0	2	0
灰巴蜗牛 *Bradyaena ravida*		0	0	0	1	0	0
合计 Sum		210	389	1450	2247	65	126
占总数百分比 Percentage		4.68%	8.67%	32.32%	50.08%	1.45%	2.81%

由表 4 可见，遗址中出土贝类多属第一至四期（即新石器时代）遗存，约占总数的 95.75%；其中大部分出自第三及第四期遗存，即以新石器时代中晚期为主。

对其他贝类各期种属比例进行计算，结果如图 2 所示。除第六期遗存中短褶矛蚌比例明显高于其他时期，各期遗存贝类种属比例大体一致，基本未受分期影响。遗址第六期遗存主要为墓葬，多见以短褶矛蚌制品随葬的现象，对此将于后文进行具体分析论述。

（二）空间分布

1. 无齿蚌亚科

对 2217 件无齿蚌亚科标本按单位进行统计，结果如表 5 所示。

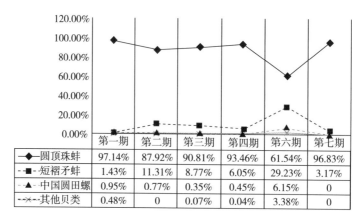

图 2 其他贝类各期种属比例
Figure 2. Proportions of shell taxa (except Anodontinae) from HTMG (with clear period).

表 5 无齿蚌亚科各单位出土情况（单位：件）
Table 5. Rates of finished product of Anodontinae from different features.

	地层 Layer			房址 House	灰沟 Ditch	灰坑 Pit	墓葬 Burial
	①层	②层	③层				
蚌料 Semi-finished product	161	183	129	327	698	603	39
成品及残件 Finished product	3	7	0	14	17	29	7
合计 Sum	483			341	715	632	46
成品率 Rate of finished product	2.07%			4.10%	2.38%	4.59%	15.22%

无齿蚌亚科标本主要出自灰沟和灰坑，约占总数的60%，在地层、房址、墓葬出土的则约占40%。

各单位类型中，灰沟出土标本数量最多而成品率较低，标本多为经初步加工的蚌料，据此推测灰沟周围是当时制作蚌器的主要场所；墓葬出土标本数量最少而成品率高，说明当时有相当比例的蚌器用于随葬；房址和灰坑为生活场所，关系较密切，标本成品率相近；散落在地层中的标本成品率最低。

2. 其他贝类

对其他贝类按单位进行统计，结果如表6所示。

表6　其他贝类各单位出土情况（单位：件）
Table 6. NISPs of mollusk taxa (except Anodontinae) from different features.

		地层 Layer			房址 House	灰坑 Pit	灰沟 Ditch	墓葬 Burial
		①层	②层	③层				
圆顶珠蚌 Unio douglasiae	左 L	173	282	152	852	1128	427	36
	右 R	151	277	161	844	1152	420	37
短褶矛蚌 Lanceolaria glayana	左 L	24	54	10	71	114	25	11
	右 R	24	54	18	74	95	24	15
河蚬 Corbicula fluminea	左 L	0	0	0	0	1	0	0
	右 R	0	0	0	1	2	0	0
虾夷盘扇贝 Mizuhopecten yessoensis		0	1	0	0	0	0	0
中国圆田螺 Cipangopaludina chinensis		88	11	4	8	14	6	4
纵肋织纹螺 Nassarius variciferus		8	0	0	0	2	1	2
灰巴蜗牛 Bradyaena ravida		1	1	0	0	0	1	0
合计 Sum		469	680	345	1850	2508	904	105
			1494					
占总数百分比 Percentage			21.78%		26.96%	36.55%	13.18%	1.53%

据统计结果进行种属比例计算，结果如图3所示。

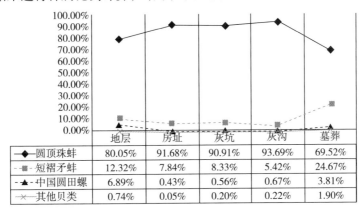

	地层	房址	灰坑	灰沟	墓葬
圆顶珠蚌	80.05%	91.68%	90.91%	93.69%	69.52%
短褶矛蚌	12.32%	7.84%	8.33%	5.42%	24.67%
中国圆田螺	6.89%	0.43%	0.56%	0.67%	3.81%
其他贝类	0.74%	0.05%	0.20%	0.22%	1.90%

图3　其他贝类各单位种属比例
Figure 3. Proportions of mollusk taxa (except Anodontinae) from different features.

圆顶珠蚌在房址、灰坑及灰沟中比例较高，而在地层和墓葬中比例偏低。此类蚌出土数量多、人工制品少，推测主要作为肉食资源被食用，房址、灰坑及灰沟中的圆顶珠蚌多为餐余垃圾。

短褶矛蚌在墓葬中比例高于其他单位。遗址中第六期墓葬普遍以短褶矛蚌制品随葬，说明此类蚌与墓葬的关联度较其他贝类大。

中国圆田螺在地层中比例高于其他单位，且出土数量少、不见人工制品，说明此种贝类没有获得当地居民关注。

其他贝类在墓葬中比例高于其他单位，说明当地居民倾向于将不常见的贝类作为随葬品。

三　贝制品研究

贝壳质地坚薄、表面光洁、取材方便、易于加工，甚至无需经过复杂的加工工序即可直接取用。在遗址中发现有叠放整齐、无加工痕迹的无齿蚌亚科蚌壳，可能是直接用作盛放食物的器皿，即与盘的使用方法大致相同。然而古代居民对贝壳的利用并不仅限于此，如遗址中发现的多样贝制品就是对贝类资源的一种重要利用方式。

（一）无齿蚌亚科制品

2217件，其中蚌料2140件、蚌器及残件77件，种类有穿孔蚌刀、蚌匙、片状蚌饰、蚌珠串饰、三棱形蚌饰、矛形蚌饰等。

无齿蚌亚科是该遗址贝制品的主要原料，推测原因有二：其一，在遗址周围水域中此蚌数量最多，是最易获得的贝类资源；其二，此蚌个体硕大，壳体较厚且坚，壳面平直，适于加工成各种蚌器。

1. 蚌料

（1）蚌壳分区

为方便记录蚌料所用的蚌壳部位，据出土情况将无齿蚌亚科壳体分为六区（图版18：图1）。

A区：蚌壳后侧和背侧三角帆部区域。壳体较薄且不平整，脆弱易碎。
B区：蚌壳腹侧区域。壳体较厚且壳面平缓，边缘平直。
C区：蚌壳前腹侧区域。壳体较厚，壳面较平缓，边缘稍有弧度。
D区：蚌壳前侧区域。壳体偏厚，壳面较平缓，边缘弧度大。
E区：蚌壳中心区域。壳体最薄，弧起程度小。
F区：蚌壳壳顶区域。壳体厚度适中，弧起程度大。

B区、C区、D区蚌料适于加工成器，是遗址中最常见的蚌料、蚌器取材部位。A区、E区、F区蚌料较少见，其中A区、E区蚌料仅能用于加工对蚌料要求不高的蚌珠，F区蚌料则因形态优势被用于加工蚌匙。

（2）分类统计

将蚌料表面生长轮脉[4]与现生标本比对确定分区。以取材区域和形状尺寸为划分标准,将2140件已完成初步定型但加工趋势不明的蚌料分为12种类型(表7),各类型典型标本见图版18:图2。

表7 无齿蚌亚科蚌料分类
Table 7. Classification of semi-finished products of Anodontinae shells.

蚌料类型 Type	取材区域 Zone	形状 Shape	长度区间 Length（cm）	宽度区间 Width（cm）	出土数量 Number
I	B＋C	边缘有弧度的长条形	10～20	3～6	389
II	B＋C/C＋D	近扇形	10～15	6～10	39
III	F区壳顶处	近椭圆形	4～10	2～8	20
IV	F区壳顶后	不规则形	6～10	5～8	17
V	C/D	纵长条形	4～7	2～5	70
VI	B	纵长条形	4～7	2～5	36
VII	C/D	近方形（小）	2～4	2～4	45
VIII	B	近方形（小）	2～4	2～4	18
IX	B	平直的长条形	3～10	3～6	624
X	C/D	近三角形	4～10	4～9	360
XI	C/D	近方形（中）	4～8	4～8	487
XII	C/D	近方形（大）	8～10	8～10	35

（3）加工工艺流程

经选料、截料、改料三步获得蚌料。

选料:选择个体硕大、壳体坚厚、壳面平直的无齿蚌亚科作为原料。

截料:根据需要截取无齿蚌亚科不同部位,绝大多数蚌料截取自蚌壳B、C、D三区,这部分的壳面不仅弧度、厚度符合大多数蚌器制作需要,而且易于截取、加工。少数蚌料截取自蚌壳E区、F区,E区蚌壳薄、弧度小,适合加工蚌珠;F区壳面厚度适中,壳顶处形成天然的弧度,符合蚌匙制作需要。截料时采用砸击法或剔刮法。

改料:方法以打制、剔刮为主。剔刮时先用锋利工具在需要分割的部位来回划动(图版19:图1),即将断裂时用手握住蚌料两端直接掰断。剔刮法截取的蚌料茬口平齐(图版19:图2),对蚌壳破坏小,能使蚌料利用率趋近最大化。

经过剔刮法处理的蚌料可继续用此法进行分割,以获得更小的蚌料。经观察,笔者认为III型、IV型、VII型及VIII型四种蚌料即是如此获得。

（4）用途推测

12种蚌料中,I型、II型、V型、VI型、IX型及X型蚌料已完成初步改料,形状尺寸合手,有蚌壳边缘形成的天然刃部,可作为简易蚌刀手持使用。由于未经精细打制和磨制定型,难以确定其加工趋势,也有可能作为其他蚌器的毛料,故将其归作蚌料更为妥当。

2. 穿孔蚌刀

（1）类型划分

出土 15 件，根据器身形状可分为 A、B 两型。

A 型　无齿蚌刀，可分为三种亚型。

AⅠ型　单孔直刃无齿蚌刀，6 件，均残。长条形，背侧及腹侧平直，腹侧为刃，背侧一孔。标本 13DHAIF4①：11（图版 19：图 3，6），从中间断开，残存一侧，边缘有缺损，方位不明。最大长 66.34 mm，最大宽 38.14 mm。背侧经磨制。孔由壳外向壳内钻制而成，壳外孔径 5.51 mm，壳内孔径 4.03 mm。

AⅡ型　单孔凸刃无齿蚌刀，5 件，均残。长条形，背侧及腹缘微弧，腹侧为刃，背侧一孔。标本 13DHAIF4①：12（图版 19：图 3，3），从中间断开，残存一侧，取自左壳。最大长 62.26 mm，最大宽 37.15 mm。背侧经打制。孔由壳内向壳外钻制而成，边缘残，壳外孔径 4.36 mm，壳内孔径 5.50 mm。

AⅢ型　双孔凸刃无齿蚌刀，2 件，长条形，背侧及腹部微弧，腹侧为刃，背侧两孔。标本 13DHAIF6①：1（图版 19：图 3，2），边缘残缺，取自左壳。最大长 129.27 mm，最大宽 39.69 mm。两孔均由壳外向壳内钻制而成。前侧孔略小，壳外孔径 5.24 mm，壳内孔径 4.39 mm；后侧孔略大，壳外孔径 6.73 mm，壳内孔径 5.47 mm。

B 型　有齿蚌刀，可分为以下两种亚型：

BⅠ型　单孔斜直刃有齿蚌刀，1 件。与 A 型蚌刀形制类似，区别为刃部呈锯齿状。标本 11DHAⅢT1309②：3（图版 19：图 3，4），较完整，方位不明。最大长 103.09 mm，最大宽 38.89 mm。孔由壳内向壳外钻制而成，壳内孔径 6.71 mm，壳外孔径 4.43 mm。

BⅡ型　单孔凸刃有齿蚌刀，1 件。标本 11DHAⅢF1：1（图版 19：图 3，5），腹缘略残，取自左壳。最大长 146.36 mm，最大宽 47.70 mm。背侧经精细打磨，后侧磨圆。刃部前侧有不明显的锯齿，应是多次使用所致。孔为对钻而成，壳外孔径 6.91 mm，壳内孔径 8.59 mm，中间最小孔径 4.52 mm。

（2）加工工艺流程

据遗址中较完整蚌刀的形态及尺寸，推测蚌刀是由 I 型蚌料（图版 19：图 3，1）加工而成。经过选料、截料、改料三个步骤获得 I 型蚌料后，还需定型、穿孔两步完成加工流程。

定型：通过打制法和磨制法修整 I 型蚌料。先将其背缘修理平整，再根据需要对腹缘进行加工。若加工无齿蚌刀，则将其薄锐腹缘作为刃，可能加以打磨使之更锋利；若加工有齿蚌刀，则将刃部打成细密锯齿状。

穿孔：根据需要进行穿孔。根据孔的形态推测加工时主要采用锥钻，有单钻、对钻两种手法。单钻时，钻孔方向不固定。

（3）功能分析

蚌刀形制受石刀影响，其功能也应与石刀大体相同，但由于材质所限，蚌刀难以对高硬度、高韧度的对象进行加工。

石刀的用途，有学者归纳总结为三种：作农具用、作刮削用、作割切用[5]。而通过对石刀表面残留物的淀粉粒和植硅体分析，也证明其主要功能之一为收割，也可能用于刮削或切割食物[6]。还有学者根据刃部形态推测判断凸刃蚌刀应是加工工具，不便用于收割，其作用是切割物品，既可作厨刀，也可作屠刀等[7]。但后套木嘎遗址刃部微凸的AⅡ、AⅢ、BⅡ型蚌刀弧度并不明显，用作农具也未尝不可。

综上，笔者认为A型蚌刀可作收割工具、屠刀和厨刀，既能胜任收割农作物的工作，又能完成对食材的处理；B型蚌刀则更侧重于收割功能，其锯齿状的刃部在收割谷穗时可以增加摩擦力，提高收割效率。使用时，将绳穿过蚌刀上的孔绑在手上或木棒上，即可进行刮、削、划、割等。

3. 蚌匙

（1）类型划分

出土17件，据器身形状可分为A、B两型。

A型 尖头蚌匙，8件，均残。总体近椭圆形，一侧略尖，中间下凹，边缘有打制痕迹。标本11DHAⅢT1512②：2（图版20：2），较完整，取自右壳。最大长69.72 mm，最大宽45.78 mm，器高14.17 mm。

B型 平头蚌匙，9件，除1件较完整外均残。总体呈圆角长方形，一端略窄，另一端略宽，中间下凹，边缘有磨圆痕迹。标本11DHAⅢT1512②：1（图版20：4），较完整，取自左壳。最大长67.98 mm，最大宽44.51 mm，器高11.12 mm。

（2）工艺流程分析

据蚌匙半成品及蚌料的形态尺寸，推测此类蚌器是由取自F区的Ⅲ型蚌料（图版20：1）和Ⅳ型蚌料（图版20：3）加工而成。经过选料、截料、改料三个步骤获得蚌料后，还需进一步定型。定型时，先用打制法把蚌料修整成型，再将打制痕迹磨平以获得光滑的边缘。

（3）功能分析

蚌匙与现在汤匙外形十分相像，功能也大致相同，是作为餐具使用的。A、B两型蚌匙功能应无区别，使用时直接用手指捏住边缘舀取食物。

4. 片状蚌饰

（1）类型划分

出土35件，据器身形状可分为A、B两型。

A型 圆形蚌片，据尺寸分成两个亚型。

AⅠ型 大圆形蚌片，12件，保存情况较好，加工程度不同。

壳外侧生长线明显，边缘稍有起伏，略呈椭圆形的有5件。标本12DHAⅢG18：2（图版21：4），较完整，方位不明。最大径84.30 mm，最小径57.16 mm。

壳外侧生长线明显，边缘稍有起伏，略呈圆形的有3件。标本11DHAⅢH92：1（图版21：5），较完整，方位不明。最大径53.08 mm，最小径50.29 mm。

壳外侧不见生长线，边缘较光滑，略呈圆形的有2件。标本11DHAⅢM30：2（图版21：6），较完整，方位不明。最大径50.05 mm，最小径49.48 mm。

壳外侧不见生长线，边缘光滑，呈圆形的有 2 件。标本 12DHAⅢG18：2（图版 21：7），较完整，方位不明。最大径 40.70 mm，最小径 37.55 mm。边缘有穿孔留下的痕迹。

AⅡ型　小圆形蚌片，7 件，可分为三种。

AⅡ1 型　无孔小型圆蚌片，5 件，保存情况较好，加工程度不同。

壳外侧未经加工，生长线明显，四周经打制，略呈圆形的有 3 件。标本 11DHAⅢH92：114（图版 21：14），较完整，方位不明。最大径 28.16 mm，最小径 27.88 mm。

壳外侧未经加工，生长线明显，四周经磨制，边缘光滑，呈圆形的有 2 件。标本 11DHAⅢT1412②：4（图版 21：15），较完整，方位不明。最大径 23.89 mm，最小径 23.49 mm。

AⅡ2 型　单孔小圆形蚌片，1 件。标本 11DHAⅢT1305②：3（图版 21：16），较完整，方位不明，壳外侧经打磨，不见生长线，四周经磨制。最大径 16.71 mm，最小径 16.38 mm。近边缘处有一由壳内向壳外钻成的孔，壳内孔径 3.99 mm，壳外孔径 2.98 mm。

AⅡ3 型　双孔小圆形蚌片，1 件。标本 11DHAⅠH92：4（图版 21：17），较完整，方位不明，壳外侧经打磨，生长线纹路较浅，四周经磨制。最大径 15.33 mm，最小径 15.05 mm。其上两孔，壳内外孔径均为 2.70 mm，难以判断钻孔方向。

B 型　璜形蚌片，16 件，保存情况较好，加工程度不同。

四周边缘稍有起伏，略呈璜形的有 3 件。标本 12DHAⅢF8：13（图版 21：2），较完整，取自右壳。最大长 85.63 mm，最大宽 36.35 mm。

四周边缘光滑，呈璜形的有 13 件。标本 13DHAⅠF6①：38（图版 21：3），较完整，取自右壳。最大长 83.02 mm，最大宽 26.76 mm。

（2）工艺流程分析

据蚌片半成品及蚌料的形态尺寸，推测此类蚌器是由Ⅸ型蚌料（图版 21：1）、Ⅹ型蚌料（图版 21：11）、Ⅺ型蚌料（图版 21：8）加工而成。经过选料、截料、改料获得以上三型蚌料后，每型蚌片有不同的加工工序。

AⅠ型蚌片还需定型、穿孔两步。

定型：对Ⅸ型蚌料边缘进行打制，获得近椭圆形（图版 21：4）和近圆形（图版 21：5）的半成品，再对蚌料边缘及蚌壳外侧进行磨制，完成定型，获得近圆形蚌片（图版 21：6）。

穿孔：根据需要在壳面穿孔，获得成品（图版 21：7）。

AⅡ型蚌片还需要二次改料、定型、穿孔、抛光四步。

二次改料：通过锯割法和剔刮法将Ⅸ型、Ⅹ型及Ⅺ型蚌料加工成呈纵长条形的Ⅵ型蚌料（图版 21：9）和Ⅴ型蚌料（图版 21：12），再以相同手法将其加工成近方形的Ⅷ型蚌料（图版 21：10）和Ⅶ型蚌料（图版 21：13）。

定型：先对近方形蚌片边缘进行打制，修理成近圆形蚌片（图版 21：14），再磨制边缘，完成定型，获得 AⅡ1 型蚌片（图版 21：15）。

穿孔：根据需要在壳面穿孔。

抛光：对蚌片表面进行抛光处理，获得 AⅡ2 型及 AⅡ3 型蚌片（图版21：16、图版21：17）。

B 型蚌片仅需经定型即可获得。

将 Ⅸ 型蚌片后侧及上下两侧的折角打掉，利用蚌壳边缘天然的弧度，获得略呈璜形的半成品（图版21：2），对蚌片边缘进行磨制加工，获得 B 型蚌片（图版21：3）。

（3）功能分析

这种蚌片应是直接作为装饰品佩戴，但无法判定具体佩戴位置。有些圆形小蚌片上有孔，应是使用线绳类的物体从孔中穿过，坠于耳部、颈部或身体其他部位；也可能是作为镶嵌物，嵌在其他物品上作为装饰。

5. 蚌珠串饰

（1）标本描述

出土 3 件（图版22：图1），出土时不见线绳，为大量大小相若、厚度不一的圆柱状带孔蚌珠，其中一些蚌珠已经残损，据形态推测为串饰，故将其穿起。3 件蚌珠串饰现存长度不一：标本 12DHAⅢM91：1 长约 75 cm，标本 12DHAⅢM92：1 长约 85 cm，标本 12DHAⅢM79：1 长约 100 cm，推测原串饰长度均长于现存。串饰主要由圆形带孔的小蚌片组成，另有鱼椎骨制成的骨珠和陶珠各 1 件。蚌珠直径约 3 mm，厚约 2 mm，孔位于蚌珠中心位置，孔径约 1 mm（图版22：图2，2）。

（2）加工工艺流程

因蚌珠过于细小，且数量巨大，若为磨制，则太过耗时耗力，操作性极低，故推测蚌珠串饰并非磨制而成。在遗址中发现有加工痕迹的标本 11DHAⅢT1005①：1（图版22：图2，1），其上有排列整齐的凹点，大小与蚌珠上的孔基本一致，凹点之间的距离也与蚌珠直径大体相同，推测为蚌珠毛料。

加工流程大致如下：

选料：选择大小合适、薄厚适宜的无齿蚌亚科作为加工对象。

打凹点：在无齿蚌亚科壳内侧打上整齐的凹点（据凹点形态推测工具为修理出尖的石器或骨器）。出土蚌珠厚度不一，故无齿蚌亚科除不平整的壳顶和帆部外，其他区域都有可能用来制作蚌珠。

骨管取珠：用较小且质地较硬的骨管（推测为鸟类肢骨），在凹点的周围摩擦转动取得蚌珠。

串珠：将取得的蚌珠串在一起（猜测使用了动物毛发或其他有机质地细线）。

（3）功能分析

此种串饰均出土于墓葬中，应是当时用于随葬的装饰品。出土蚌珠串饰的三座墓葬已被扰乱，所有蚌珠散乱分布，难以确定具体佩戴方式。猜测有三种可能：一是佩戴于颈部，作为项饰；二是缠绕于头部，作为发饰；三是缠绕于手上，作为腕饰。

6. 三棱形蚌饰

（1）标本描述

出土 1 件。标本 11DHAⅢH85：1（图版22：图3，2），较完整，方位不明，整体呈三棱形，微弧，中间较宽，两端略尖，一端有条状凹槽，侧面磨制光滑。最大长50.88 mm，最大宽9.06 mm，厚6.99 mm。

（2）加工工艺流程

该蚌器由Ⅳ型蚌料（图版22：图3，1）加工而成。经选料、截料、改料获得Ⅳ型蚌料后进行定型。采用打制法将Ⅳ型蚌料上多余部分打掉，磨制侧面，在一端锯割出凹槽。

（3）功能分析

推测使用线绳类物体绑在凹槽处，坠于身上作为装饰品。

7. 矛形蚌坠饰

（1）标本描述

出土1件，稍残，方位不明。标本12DHAⅢM90：2（图版23：图1，3），表面被土质附着物腐蚀严重。呈中间宽、两端窄的矛形，其中一端有亚腰形柄，另一端有一对钻而成的孔，已残。加工比较细致，线条流畅，边缘平滑。最大长97.72 mm，最大宽17.70 mm，厚8.98 mm，孔径约5.24 mm。

（2）加工工艺流程

据厚度推测是由Ⅻ型蚌料（图版23：图1，1）加工而成。经选料、截料、改料获得Ⅻ型蚌料后，再经二次改料、定型、穿孔、抛光四步完成加工流程。

二次改料：采用锯割法和剔刮法对Ⅻ型蚌料进行加工，获得长方形厚料块（出土此种毛料4件，均残，方位不明）。标本11DHAⅢT1405②：2（图版23：图1，2），四个侧面中三面经剔刮，一面残断。最大长65.48 mm，最大宽21.19 mm，最大厚11.13 mm。

定型：对毛料进行初步定型。主要使用打制法将多余部分修理掉，再进行精细磨制使其基本成型。

穿孔：在坠饰一端使用锥类工具以对钻法穿孔。

抛光：整体抛光，使器身线条更加流畅，表面更加光滑，形态更满足审美需求。

（3）功能分析

推测将线绳类物体穿过孔或系在柄部，坠于身上作为装饰品。

（二）短褶矛蚌制品

短褶矛蚌制品有34件，种类有穿孔蚌刀、刻划器、磨制蚌片等。

选用短褶矛蚌加工成器，推测是因其个体大小适中，且壳体细长、匀称，可直接依托自身形态进行加工，无须改料。

1. 穿孔蚌刀

（1）类型划分

出土31件，根据穿孔位置可分为A、B两型。

A型 3件，左壳2件，右壳1件，均残。在距壳前缘约3 cm处的背侧穿孔。标本11DHAⅢT1205③：34（图版23：图2，1），壳后缘略残，为右壳。最大长120.46 mm，最大宽27.94 mm，壳高12.70 mm。孔为从壳外向壳内砸制而成，孔径4.75 mm。

B 型　前侧穿孔。根据具体穿孔位置可分为两个亚型。

BI 型　24 件，左壳 11 件，右壳 13 件，保存情况较好。在前闭肌痕处穿孔，壳后侧多经磨制。标本 11DHAIIIM58：1（图版 23：图 2，2），表面风化严重，从中间断开后黏合，为右壳。最大长 101.71 mm，最大宽 20.41 mm，壳高 6.23 mm。

BII 型　4 件，左壳 1 件，右壳 3 件，保存情况较好。在壳前缘中间处穿孔，壳体前后两端均经磨制。标本 11DHAIIIM16：14（图版 23：图 2，3），较完整，为右壳。最大长 92.40 mm，最大宽 17.89 mm，壳高 6.49 mm。

（2）加工工艺流程

就蚌壳上所能观察到的加工痕迹来看，加工工艺流程大致分为选料、定型、穿孔三步。

选料：选用大小适宜的短褶矛蚌作为原料。

定型：对蚌壳腹侧后端进行磨制，修理出精致的刃部，再把蚌壳前端磨平。

穿孔：根据不同需要，以不同方式在壳面不同位置穿孔。

A 型穿孔蚌刀均采用直接打制的方式进行穿孔。使用小型带尖工具直接进行打制，留下的孔较粗糙，边缘有打制疤痕（图版 24：图 1，1）。

B 型穿孔蚌刀采用三种穿孔方式：

其一，预制台面后砸击。蚌壳前闭肌痕对应的壳体薄于其他部位，穿孔时先将壳外前闭肌痕对应位置磨平，形成椭圆形台面，再用工具在壳体最薄处轻轻砸击。这种方式留下的孔边缘薄，不规整，近椭圆形（图版 24：图 1，2）。

其二，锯割。在蚌壳前闭肌痕处，使用边缘薄锐的工具从平行于壳体、垂直于壳体两个方向锯割出孔。这种方式留下的孔边缘不规整，孔附近壳体有明显折角（图版 24：图 1，3）。

其三，管钻。均见于 BII 型穿孔蚌刀。使用中空管状工具（如鸟类肢骨制成的骨管）在壳前缘中间进行钻孔。这种方式留下的孔边缘规整，呈圆形（图版 24：图 1，4）。

（3）功能分析

穿孔蚌刀多出于遗址第六期的墓葬，属汉书二期文化遗存，形状尺寸与共出的铜刀极为相近，从扰乱轻微的墓葬中可以发现，其出土位置有一定的倾向性，往往位于墓主腰部附近。对于其用途有两种推测：其一，是以线绳穿孔佩于腰间的实用器，将绳绑缚在手上用于割划、叉取食物等，与今日之餐刀功能类似；其二，是特制的明器，下葬时以形制相近的蚌刀代替铜刀佩于死者腰际，可能是当地青铜时代早期一种特有的丧葬习俗。

2. 刻划器

（1）标本描述

出土 1 件。标本 11DHAIIIH89：B15（图版 23：图 2，4），取自右壳，利用蚌壳自身形态，仅将底部修理出短尖。最大长 77.31 mm，最大宽 14.52 mm，壳高 4.81 mm。

（2）加工工艺流程

经选料、定型两步完成。取大小适宜的短褶矛蚌，将壳体后侧两边各磨出一大小适宜的凹口，形成刻划尖。

(3) 功能分析

因蚌壳体薄易损，蚌制刻划器并不适宜加工硬度过大或韧性过强的材料。笔者认为这件刻划器可能用于加工陶器表面的刻划纹。

3. 磨制蚌片

出土2件，均残，取自左壳，为磨平蚌壳铰合齿的蚌料。因残损严重难以确定其用途，推测是饰品。标本13DHAIH47：25（图版23：图2，5），最大长40.80 mm，最大宽17.62 mm，残高4.93 mm。

(三) 圆顶珠蚌制品

1. 标本描述

出土9件，均为壳顶穿孔的蚌壳。除穿孔外未经其他加工，有直接打制、预制台面后砸击、锯割三种穿孔手法。

直接打制穿孔　6件，左壳3件，右壳3件。标本11DHAⅢH91：10（图版24：图2，1），边缘残损，为右壳。最大长46.79 mm，最大宽27.44 mm，壳高11.04 mm。壳顶孔为边缘不规整的圆形，孔周围壳面有分层脱落的现象，孔径约6.34 mm。

预制台面后砸击穿孔　2件，均为右壳。标本11DHAⅢT1312③：3（图版24：图2，2），较完整。最大长69.82 mm，最大宽32.15 mm，壳高13.44 mm。壳顶孔呈椭圆形，最大径4.77 mm，最小径3.84 mm。

锯割穿孔　1件。标本11DHAⅢH92：213（图版24：图2，3），边缘残缺严重，为右壳。最大长43.11 mm，最大宽19.50 mm，壳高8.16 mm。通体有烧痕。从侧面看孔周围壳体呈垂直缺口。孔最大径3.83 mm，最小径2.89 mm。

2. 加工工艺流程

经选料、穿孔两步完成。取大小适宜的圆顶珠蚌直接在壳顶穿孔。加工手法与短褶矛蚌蚌刀穿孔方式相同，在此不赘述。

3. 功能分析

推测两种用途：其一，作为小型蚌刀使用，利用天然锋利的腹缘为刃，使用时将线绳绑在手上，用以割划肉类等柔软物体；其二，作为装饰品，将线绳穿孔而过坠于身上。

(四) 虾夷盘扇贝制品

仅出土1件。此种贝类是冷水性种，原产于日本海周边，我国20世纪80年代将其从日本引进北部沿海，已经成大规模养殖[8]。此贝可能是古代日本海沿岸地区迁至本地的居民带来的，也可能是通过多次跨地区交换获得的。

1. 标本描述

标本12DHAⅢT1305②：7（图版24：图3），残损，由虾夷盘扇贝左壳制成，近圆形，壳质比较坚硬，一角被土壤锈蚀以致缺失，壳内有碱质附着物。该标本有两处破损，一处位于壳面，为一不规则孔洞，推测是人为加工形成；另一处位于边缘处，露出壳质，为晚近形成。最长径为 130.15 mm，最短径为 125.58 mm，厚 2.41 mm。

2. 工艺流程分析

推测加工流程有定型、穿孔两步。

定型：将虾夷盘扇贝原有的前后两耳打掉并磨圆，修理出把手以便于使用。

穿孔：在壳面上以砸击的方式进行穿孔。

3. 功能分析

推测为蚌锄一类的农具，用于翻土、锄地之类的农业活动。可直接将手指穿过其上孔洞使用，也可用绳绑缚在木棍或粗壮树枝上使用。

四 结语

本文通过对 2011～2015 年后套木嘎遗址出土贝类遗存时空分布及贝类制品的研究，对该地贝类资源利用情况有了初步认识。

由贝类遗存时空分布的分析结果，可得出以下几点结论：

1）随时间发展，贝类制品在加工过程中产生的废料越来越少，加工工艺日趋成熟。

2）遗址中贝类种属及比例未随分期发生明显变化，说明周围自然环境比较稳定。

不同遗迹类型中贝类种属比例不同，说明古代人类会有意识地根据需求选择相应贝类进行利用。如圆顶珠蚌主要用于食用，很少加工成器；短褶矛蚌食用比例小，制成蚌器的比例高；无齿蚌亚科则在食用和加工成器两方面都占有很大比重。

从贝类制品的制作、使用情况来看，当地原始居民已熟练掌握打制、磨制、锯割、剔刮、穿孔等技术，并将这些技术运用到工具及装饰品制作中。可见原始居民已具备较成熟的工具加工能力，并可进行大量的标准化制作，他们会有意识地使用不同蚌料制作不同器物，对各类器物的形状、大小有明确的判断力，同时具备审美情趣，能制作精美的装饰品。

该遗址的贝类制品有以下特点：

1）不同种属的贝类制品加工流程不同。如无齿蚌亚科制品的加工工序复杂烦琐，而其他贝类制品加工工序相对简单。

2）原始居民根据不同贝类的形态特点加工蚌器，充分利用贝壳各个分区，取料目的明确，加工技术纯熟，多样的穿孔技术尤为突出。

3）遗址内蚌料利用具有复杂性，同一类型的蚌料既可作为工具直接使用，也可以进行二次改料，加工成更精细的蚌器，充分体现了古代居民"因料制宜"的智慧。

以上信息对了解后套木嘎遗址居民的工具制作、手工业发展水平以及生活场景等提供了帮助，可更进一步探寻该遗址居民生产生活的整体面貌。

注释

[1] 王立新、霍东峰、石晓轩、史宝琳:《吉林大安后套木嘎遗址发掘取得重要收获》,《中国文物报》2012年8月17日第8版。

[2] 吉林大学边疆考古研究中心、吉林省文物考古研究所:《吉林大安后套木嘎遗址AⅢ发掘简报》,《考古》2016年第9期,第3~24页。

[3] 栾伊婷:《后套木嘎遗址古代牛的分子考古学研究》,吉林大学硕士学位论文,2016年。

[4] 生长轮脉指贝壳外面以壳顶为中心,与腹缘平行呈同心圆排列的生长线,该生长线反映瓣鳃类的年龄。参见刘月英、张文珍、王跃先、王恩义:《中国经济动物志——淡水软体动物》,科学出版社,1979年。

[5] 饶惠:《略论长方形有孔石刀》,《考古通讯》1958年第5期,第40~45页。

[6] 马志坤、李泉、郇秀佳、杨晓燕、郑景云、叶茂林:《青海民和喇家遗址石刀功能分析:来自石刀表层残留物的植物微体遗存证据》,《科学通报》2014年第13期,第1242~1248页。

[7] 王仁湘:《论我国新石器时代的蚌制生产工具》,《农业考古》1987年第1期,第145~155页。

[8] 齐钟彦:《中国经济软体动物》,中国农业出版社,1996年。

辉县孙村遗址殷墟文化时期动物骨骼的稳定同位素分析[*]

Stable Isotopic Analysis of Animal Bones from the Suncun Site, Huixian, Henan, China during the Yinxu Culture (ca. 1250 – 1046 BC)

侯亮亮[1] 张国硕[2] 戴玲玲[3] 侯彦峰[4]

Liangliang Hou[1], Guoshuo Zhang[2], Lingling Dai[3], Yanfeng Hou[4]

1. 山西大学历史文化学院，太原
2. 郑州大学历史学院，郑州
3. 辽宁师范大学历史文化旅游学院，大连
4. 河南省文物考古研究院，郑州
1. Shanxi University, Taiyuan, China
2. Zhengzhou University, Zhengzhou, China
3. Liaoning Normal University, Dalian, China
4. Henan Provincial Institute of Cultural Heritage and Archaeology, Zhengzhou, China

摘要：随着中国北方新石器时代中晚期粟黍农业的发展、强化及扩张，粟黍及其副产品在先民、家猪和家犬食物结构中的比重越来越大，使得他们的食物结构表现出极强的相关性，进而可将家猪和家犬的食物结构作为重建同时期临近区域先民生业经济的替代性指标。然而随着社会的复杂化，外来及新的农作物种植和推广，家畜饲喂和管理更加专业化，家猪、家犬与先民的食物结构是否会有相应的变化，是否还存在密切的相关性，是否还可以将家猪和家犬的食物结构作为重建先民生业经济的替代性指标？本文对河南辉县孙村遗址殷墟文化时期兽骨进行了 C、N 稳定同位素分析，并比较同时期临近的殷墟遗址已发表先民、家猪和家犬的稳定同位素数据。结果显示，家猪和家犬的食物结构不能作为重建殷商文化时期先民生业经济的替代性指标，但可以作为大部分先民生业经济的间接、笼统的指示参考。

关键词：猪和狗；替代性指标；孙村遗址；C、N 稳定同位素

Abstract: To reconstruct subsistence strategies of past populations, human remains are

---

[*] 本研究由教育部人文社会科学研究青年基金（批准号：15YJC780003）资助。

analyzed by testing stable isotope ratios of carbon ($\delta^{13}C$) and nitrogen ($\delta^{15}N$). Unfortunately, human skeletons are not always accessible at archaeological sites. In such cases, domestic animals (especially dogs and pigs) are taken as the parameter of human diets. This is based on the assumption that these domestic animals had a commensal relationship and shared similar food with humans. Thus, analyzing the stable isotopic ratios of domestic animals, to some extent, allows to reconstruct human diets.

Millets were widely cultivated and consumed as the staple food for humans in North China from the beginning of the Neolithic period till the Three Dynasties (Xia, Shang, and Zhou). Meanwhile, livestock, especially domestic pigs and dogs, were raised in large numbers as the most important meat resources for ancient populations. Based on the intensification and expansion of millet-based agriculture, ancient humans, domestic pigs and dogs relied primarily on the millet-based diet in North China during the Middle and Late Neolithic periods, which had been confirmed by their similar stable isotopic values in previous studies.

Since the Late Neolithic period, wheat and barley had been introduced into North China. The diets of humans and domestic pigs and dogs gradually changed from C_4 - based to C_3 and C_4 mixed. Literature and archaeological materials indicate that wheat and barley were important crops during the Shang Dynasty (1600 – 1046 BC) in the Central Plains. In this paper, we select domestic animal bones from Suncun, a Yinxu Culture (ca. 1250 – 1046 BC) site to test whether the stable isotope signatures of pigs and dogs could be the parameter of the human diets.

The site of Suncun, located in the southwest of the Suncun Village, Gaoxiang Town, Huixian City, Henan Province, China, is sitting at the foot of the Taihang Mountains. In order to protect the cultural heritage in the region of the "South-North Water Transfer Project" midline, an area of 2020 m² at Suncun was excavated from July to October in 2006, when the archaeological features such as pits, hearths and ditches were revealed. Most cultural materials found from Suncun were dated to the Proto-Shang Culture to the Han Dynasty, and can be divided into four periods: the Proto-Shang Culture (ca. 2000 – 1600 BC), the Yinxu Culture (ca. 1250 – 1046 BC), the Warring States (475 – 221 BC) and the Han Dynasty (206 BC – AD 220). The cultural materials of the Yinxu Culture were abundant. Although no burials were found at Suncun, lots of animal bones, including pigs, dogs, cattle, sheep, and deer were recovered on-site.

Totally 26 animal bones from Suncun were selected for stable carbon and nitrogen isotopes analysis. Results show that the $\delta^{13}C$ and $\delta^{15}N$ values of pigs [(−6.9 ± 1.1)‰, (6.7 ± 0.3)‰, N = 8], dogs [(−6.4 ± 0.5)‰, (7.7 ± 0.7)‰, N = 6], and cattle [(−9.3 ± 0.9)‰, (6.5 ± 1.1)‰, N = 6] are generally higher than those of deer [(−20.3 ± 0.9)‰, (5.6 ± 0.9)‰, N = 4], indicating that the former three animal species relied primarily on C_4-based food (millets), whereas the deer relied on C_3-based food (wild plants). Meanwhile, results of two sheep/goats (-16.3‰, 6.3‰, N = 1; -10.3‰, 7.1‰, N = 1) show different stable isotope values, which could be related to their feeding habits. Different stable isotopic patterns were recognized by comparing δ^{13}C and δ^{15}N values of pigs, dogs, and humans from Suncun and Yinxu, a contemporary site near Suncun. This suggests that social and dietary complexities during the Yinxu Culture period already had effects on the dies of domestic pigs and dogs, which were not available to be the parameter to discuss human diets and subsistence.

Keywords: pig and dog, proxy, Suncun, stable carbon and nitrogen isotopes

一 引言

更新世晚期至全新世初期，粟黍（C_4植物）开始被我国先民驯化和栽培[1]，几乎同时，猪和狗也开始被先民驯化和饲喂[2]。其后，以二者为基础的农业经济体系不仅奠定了我国北方地区新石器时代先民生业经济的传统和模式，而且有效推动了该地区长期文化的繁盛和人口的显著增长[3]。

相关研究表明，随着北方地区粟黍农业的发展和扩张，先民越来越依赖粟黍及其副产品，具体表现在粟黍在其食物结构中的比重越来越大[4]。与此相伴，先民也开始越来越多的用粟黍及其副产品来饲喂家猪和家犬[5]。这种依赖粟黍农业的生产体系，使得先民及家猪、家犬的食物结构表现出极强的相关性[6]。

近年来，人和动物硬组织（骨骼和/或牙齿）的稳定同位素分析成为观察这种相关性的有效手段，可以真实地反映人和动物长期的摄食特征和饮食习惯[7]。随着我国北方地区不同时期人和动物硬组织稳定同位素数据的积累和丰富，稳定同位素视角下该地区先民及家猪、家犬食物结构相关性的轮廓日益清晰和明确[8]。初步的研究成果表明，在一定的时空框架下，先民和家猪、家犬的食物结构极为相似，甚至没有差异[9]。这使我们有机会从重建先民生业经济替代性指标的角度去考虑相关问题，即当考古遗址中缺失人骨或其他情况下不能获取人的硬组织时，可以将家猪和家犬的食物结构作为先民生业经济的重要参考[10]。

龙山晚期至三代（夏、商、周），社会的复杂化程度加剧，即在孕育早期国家和文明的同时，先民的阶层划分日益明确，先民的生业经济开始发生较大的变化[11]。首先，随着粮食作物种类的多样化，特别是外来农作物的推广和种植，中国北方粟黍农业体系逐步受到冲击[12]。相关研究成果表明，小麦、大豆及水稻等（C_3植物）开始被我国北方先民栽培和食用，并对不同时空下先民的食物结构和生业经济产生相应的影响[13]。其次，家畜饲养业得到了极大的发展，先民广泛饲喂马、牛、羊、鸡、狗、猪等六畜，对家畜的控制和管理日益专业化，特别是对家畜的二次开发和利用[14]。最后，不同阶层的人群对动植物利用的差异逐步明显[15]。因此，观察和研究这一阶段家猪和家犬的食物结构，一方面可以了解以上变化对家畜食物结构的影响及对先民饲喂家畜的行为影响；另一方面还可以确定先民与家猪、家犬的食物结构是否依然有较强的相关性，进而判断是否可以继续将家猪和家犬的食物结构作为重建先民生业经济的替代性指标。

鉴于此，本文尝试用 C、N 稳定同位素分析的方法，研究河南辉县孙村遗址殷墟文化时期（ca. 1250～1046 BC，相当于商代中晚期）的动物骨骼，特别是家猪和家犬的骨骼，重建它们的食物结构和饲养方式。同时与同时期临近的殷墟遗址先民的稳定同位素数据进行比较，观察其食物结构的相关性，验证家猪和家犬的食物结构是否可以作为重建先民生业经济的替代性指标，以期对生业考古研究提供借鉴。

二　材料与方法

（一）遗址背景

辉县孙村遗址位于河南省辉县市高庄乡孙村西南，太行山脉东南。为配合"南水北调"中线的文物保护工作，2006 年 7 月至 10 月，郑州大学历史学院、新乡市文物局、辉县市文物局等单位对该遗址进行了抢救性发掘[16]。

此次发掘面积为 2020 平方米，发现了较为丰富的先商文化、殷墟文化、战国至王莽时期的遗存和遗物。其中殷墟文化时期的遗存和遗物最为丰富，特别是发现较多动物骨骼，以猪、牛、羊、鹿等动物的骨骼为主[17]。遗憾的是该遗址无人骨样品发现。

（二）样品选择

实验所用骨样均采自孙村遗址。动物骨样包括猪 8 例、狗 6 例、牛 6 例、羊 2 例、鹿 4 例，共 26 例。样品的详细情况如表 1 所示。

（三）骨骼胶原蛋白制备

样品的处理主要参考 Richards 和 Hedges 的方法[18]，并略作修改。机械去除骨样内外表面的污染物，称取 2 克左右，用 0.5mol/L HCl 溶液 4℃下浸泡，每隔两天更换酸液，直至骨样松软，无明显气泡，用去离子水清洗至中性。再用 0.125mol/L NaOH 溶液浸泡 20 小时，去离子水洗至中性，浸于 0.001mol/L HCl 溶液在 70℃下加热 48 小时后趁热过滤，−20℃冷冻。冷冻干燥 48 小时后得骨胶原，称重，计算骨胶原得率（骨胶原重量/骨样重量）（表1）。

表 1 孙村遗址动物骨的种属、出土单位及各项测试数据
Table 1. Archaeological background and the test data of human bones from Suncun.

编号	单位	种属	$\delta^{13}C$ ‰	$\delta^{15}N$ ‰	C %	N %	C/N	骨胶原得率（%）
P1	2006HSIH11:1	猪（Sus scrofa domesticus）	-6.5	6.3	25.5	9.4	3.2	12.4
P2	2006HSIH11:8	猪（Sus scrofa domesticus）	-6.9	6.9	33.5	12.1	3.2	9.7
P3	2006HSIH50:9	猪（Sus scrofa domesticus）	-6.6	7.0	34.9	13.5	3.0	8.6
P4	2006HSIH8:3	猪（Sus scrofa domesticus）	-6.9	6.3	30.1	12.3	2.9	9.5
P5	2006HSIH50:10	猪（Sus scrofa domesticus）	-9.7	6.5	42.8	15.7	3.2	7.4
P6	2006HSIH43:2	猪（Sus scrofa domesticus）	-6.2	6.9	44.0	15.8	3.2	7.4
P7	2006HSIH50:2	猪（Sus scrofa domesticus）	-6.4	6.8	40.6	15.0	3.2	7.8
P8	2006HSIH40:3	猪（Sus scrofa domesticus）	-6.3	7.0	37.4	14.2	3.1	8.2
D1	2006HSIH50:17	狗（Canis familiaris）	-6.7	9.0	44.1	16.1	3.2	7.2
D2	2006HSIH27:13	狗（Canis familiaris）	-5.9	7.8	41.7	15.4	3.2	7.6
D3	2006HSIH27:23	狗（Canis familiaris）	-6.3	7.6	43.1	15.8	3.2	7.4
D4	2006HSIH4:2	狗（Canis familiaris）	-6.3	7.2	41.5	15.2	3.2	7.7
D5	2006HSIH67:2	狗（Canis familiaris）	-6.1	7.5	41.8	15.5	3.1	7.5
D6	2006HSIH33:2	狗（Canis familiaris）	-7.3	7.2	35.7	13.7	3.0	8.5
C1	2006HSIH11:2	牛（Bos taurus）	-9.5	6.9	25.9	9.5	3.2	12.3
C2	2006HSIH50:14	牛（Bos taurus）	-9.5	6.9	40.6	15.0	3.1	7.8
C3	2006HSIH25:21	牛（Bos taurus）	-10.5	8.1	38.0	14.2	3.2	8.3
C4	2006HSIH50:3	牛（Bos taurus）	-9.5	6.8	40.7	15.0	3.2	7.8
C5	2006HSIH11:3	牛（Bos taurus）	-8.0	5.4	37.0	13.9	3.1	8.4
C6	2006HSIH11:12	牛（Bos taurus）	-8.5	5.1	43.6	15.8	3.2	7.4
S1	2006HSIH131:2	羊（Caprinae）	-10.3	6.3	24.0	9.4	3.0	12.4
S2	2006HSIH14:2	羊（Caprinae）	-16.3	7.1	33.0	12.0	3.2	9.7
DE1	2006HSIH43:11	鹿（Cervidae）	-20.6	6.9	38.7	14.6	3.1	8.0
DE2	2006HSIH114:3	鹿（Cervidae）	-20.9	5.2	43.1	15.8	3.2	7.4
DE3	2006HSIH33:5	鹿（Cervidae）	-19.0	5.6	36.9	14.1	3.0	8.3
DE4	2006HSIH25:38	鹿（Cervidae）	-20.6	4.8	42.7	15.5	3.2	7.5

（四）C、N 稳定同位素测试及分析

骨骼胶原蛋白中 C、N 元素含量及其稳定同位素比值的测定，在中国农业科学院农业环境与可持续发展研究所测试中心进行。取少量骨胶原，称重，使用 Elementar Vario-Isoprime100 型稳定同位素质谱分析仪（Isoprime 100 IRMS coupled with Elementar Vario）测试其 C、N 含量及同位素比值。测试 C、N 含量所用的标准物质为磺胺（Sulfanilamide）。C、N 稳定同位素比值分别以 USGS 24 标定碳钢瓶气（以 PDB 为基准）和 IEAE-N-1 标定氮钢瓶气（以 AIR 为基准）为标准，每测试 10 个样品插入一个实验室自制胶原蛋白标样[δ^{13}C 均值为 (-14.7 ± 0.2)‰，δ^{15}N 均值为 (6.9 ± 0.2)‰]。分析精度都为 ±0.2‰，测试结果以 δ^{13}C（相对于 PDB）、δ^{15}N（相对于 AIR）表示，详见表 1。

运用 SPSS 15.0 和 Origin 8.0 软件对所测数据进行统计、分析。

三　结果与讨论

（一）骨骼的污染判定

骨骼在埋藏过程中受到温度、湿度及微生物等因素的影响，其结构和化学组成将发生改变[19]。判断骨样是否被污染，是进行 C、N 稳定同位素分析的前提。由表 1 可知，所有样品的骨胶原提取率在 7.2% ~ 12.4% 之间，均值为 (8.5 ± 1.6)%（N = 26），和现代样品（约含 20% 骨胶原）有较大差距，表明骨胶原在埋藏过程已发生不同程度的降解[20]。判断骨胶原是否污染的最重要指标当属骨胶原的 C、N 含量和 C/N 摩尔比值。样品 C、N 含量在 24.0% ~ 44.1% 和 9.4% ~ 16.1%，均值分别为 (37.7 ± 6.0)%（N = 26）和 (14.0 ± 2.0)%（N = 26），接近于现代样品（C、N 含量分别为 41%、15%）[21]；样品的 C/N 摩尔比值都在 2.9 ~ 3.2 之间，也落于未受污染样品的范围内（2.9 ~ 3.6）[22]。由此判定，全部样品提取出的骨胶原均可用作稳定同位素分析。

（二）动物的食物结构

图 1 为孙村遗址 5 种动物——猪、狗、牛、羊和鹿的 δ^{13}C、δ^{15}N 散点图。由图可见，食草类动物中鹿的数据相对集中，说明鹿的食物结构较为一致；而食草类动物中牛和羊的数据则比较分散，存在较大差异；杂食类动物猪和狗的数据也相对集中，表明两种动物的食物来源基本相似。

鹿的数据相对集中，其 δ^{13}C 值和 δ^{15}N 值分别集中分布于 -20.9‰ ~ -19.0‰ 和 4.8‰ ~ 6.9‰ 之间，均值分别为 (-20.3 ± 0.9)‰（N = 4）和 (5.6 ± 0.9)‰（N = 4），表明它们主要以 C_3 类植物为食，符合食草动物的食性特征[23]。此外，相关研究表明，一个地区典型的野生动物食谱很大程度上可以反映出当地的野生植被情况[24]，据此判断辉县孙村殷墟文化时期野生植被是以 C_3 类植物（野生草本植物）为主。

尽管羊和牛也皆属食草类动物，但与鹿的食物来源存在较大的差异。首先，羊的

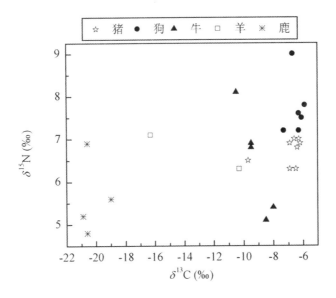

图 1 动物的 $\delta^{13}C$、$\delta^{15}N$ 散点图
Figure 1. Scatter plot of $\delta^{13}C$ and $\delta^{15}N$ of animal bones from Suncun.

数量较少，仅有两例，$\delta^{13}C$ 值差异较大，分别为 −16.3‰ 和 −10.3‰，说明食物来源存在较大的差异，可能与其属不同羊类（绵羊和山羊）有关。羊的 $\delta^{15}N$ 值均较为正常，分别为 6.3‰ 和 7.1‰，体现出食草类动物的同位素特征。牛的 $\delta^{13}C$ 值比较集中，大都分布在 −10.5‰ ~ −8.0‰ 之间，均值为 (−9.3 ± 0.9)‰（N = 6），显现出明显 C_4 类特征，表明其食物以 C_4 类植物为主；其 $\delta^{15}N$ 值尽管有所波动（5.1‰ ~ 8.1‰），但平均值 (6.5 ± 1.1)‰（N = 6）表现出食草动物的典型特征。

作为杂食类动物的猪和狗，与以上食草类动物的食物结构存在一定的差异。由图 1 可以看出，猪、狗的数据分别相对集中，表明它们各自的食物结构较为一致。猪的 $\delta^{13}C$ 值和 $\delta^{15}N$ 值分别集中分布于 −9.7‰ ~ −6.2‰ 和 6.3‰ ~ 7.0‰ 之间，均值分别为 (−6.9 ± 1.1)‰（N = 8）和 (6.7 ± 0.3)‰（N = 8），表明它们主要以 C_4 类食物，即粟黍及其副产品为食，符合杂食动物的食性特征。狗的 $\delta^{13}C$ 值与猪的类似，分布于 −7.3‰ ~ −5.9‰ 之间，均值为 (−6.4 ± 0.5)‰（N = 6），说明狗与猪的植物类食物来源基本一致。但狗的 $\delta^{15}N$ 值分布范围 7.2‰ ~ 9.0‰ 以及均值 (7.7 ± 0.7)‰（N = 6）都要略高于猪的相应值，表明狗的食物中包含更多的动物蛋白，这与狗的生活习性相符合。

（三）动物的饲喂方式

历史文献和考古发现均表明，殷商文化时期先民主要依赖传统的粟黍农业，但也开始栽培和种植小麦、大豆及水稻等（C_3 植物）[25]。如上文分析显示，辉县孙村殷墟文化时期家猪和家犬的食物结构几乎全部依赖粟黍及副产品，表明它们与粟黍农业密切相关。这与北方地区大多数新石器时期家猪和家犬的饲喂模式类似，表明殷墟文化时期先民保持了以粟黍农业为基础饲喂家猪和家犬的传统模式，小麦、大豆及水稻等（C_3植物）的栽培和种植似乎没有对殷商文化时期辉县孙村遗址家猪和家犬的饲喂产生影响。

距今4500年前后，牛、羊开始被我国先民饲喂，并形成了两种完全不同的饲喂方式，即牛主要以粟黍农业的副产品来喂养，羊则主要在野外放养。如陶寺遗址牛的C同位素分析显示先民采用了较多粟作农业的副产品来饲喂[26]。辉县孙村殷墟文化时期牛的食物结构包含了大量的粟黍和/或其副产品，说明先民主要以粟黍农业的副产品来喂养。羊的数量仅有两例，二者稳定同位素值差异较大，分别以C_3类植物和C_4类植物为主，表明不同个体的食物来源十分广泛，说明先民对其饲养方式多种多样，可能既有以野外植物为食的放养模式，也有以粟黍农业副产品为食的喂养模式。

中国的古代文献中经常有关于鹿的记载，如《诗经·大雅》"王在灵囿，麀鹿攸伏"[27]，但是人类饲喂和管理鹿的记载鲜见。殷墟文化时期辉县孙村遗址先民似乎也没有类似的行为。如上文分析显示，辉县孙村遗址鹿的食物结构主要以野生草本植物为主，显示它们很少受到人工行为的干预。辉县孙村遗址内发现较多鹿的骨骼[28]，应是先民狩猎所获。

（四）家猪、家犬与先民食物结构的关系

通过上文分析重建辉县孙村遗址殷墟文化动物群的食物结构，可以发现家猪和家犬的食物结构与粟黍农业密切相关。由于辉县孙村遗址无同一时期人骨的发现，因此无法直接用稳定同位素的方法重建先民的食物结构，也无法进一步研究先民及家猪、家犬食物结构的相关性。所幸与辉县孙村遗址相邻的殷墟地区发现了大量的人骨并开展了稳定同位素的工作，为分析和研究先民及家猪、家犬食物结构的相关性提供了重要的数据。

2003年，张雪莲对殷墟遗址出土的39例人骨进行了C稳定同位素分析，并对1例人骨进行了N稳定同位素分析，发现殷墟遗址大部分先民主要以粟黍为食物，但也有1例个体主要以C_3食物（小麦、水稻、大豆等）为食[29]。2013年，司艺对殷墟孝民屯遗址4例人骨进行C、N稳定同位素分析，发现尽管先民以粟黍为食物，但也有一定比例的C_3食物[30]。2017年，Cheung等对殷墟遗址及周边同一时期遗址更多的人骨进行了C、N稳定同位素分析，发现绝大部分先民主要以粟黍为食物，但也有一定量的C_3食物[31]。此外，Cheung等还发现，相较殷墟遗址墓主人，人牲和人殉的食物结构中粟黍所占的比例更大、肉食资源所占比例更小，似乎说明不同阶层的人群对C_3食物和肉食资源的占有存在差异[32]。2010年和2013年，闫灵通和司艺分别重建了殷墟遗址大量家猪和家犬的食物结构，发现它们也主要以粟黍为食[33]。2017年，Cheung等对殷墟遗址的鼠进行了C、N稳定同位素分析，发现鼠的食物结构中C_3食物的比重较家猪和家犬高[34]。（表2）

表2 孙村遗址及殷墟遗址人、家猪、家犬及鼠的C、N稳定同位素均值

Table 2. Average values of δ^{13}C and δ^{15}N of humans, pigs, dogs and small rodents from Suncun and Yinxu.

骨骼的种属及出土背景	个体数	δ^{13}C ± SD (‰)	δ^{15}N ± SD (‰)	数据来源
猪（辉县孙村遗址，灰坑和地层）	8	-6.9 ± 1.1	6.7 ± 0.3	本文表1
狗（辉县孙村遗址，灰坑和地层）	6	-6.4 ± 0.5	7.7 ± 0.7	
猪（殷墟孝民屯遗址，灰坑）	24	-8.0 ± 1.4	8.0 ± 0.6	闫灵通：《稳定同位素在陆生动物矿化组织研究中的应用》，中国科学院研究生院博士学位论文，2010年。
狗（殷墟孝民屯遗址，灰坑）	10	-7.8 ± 1.0	8.1 ± 1.0	
猪（殷墟铁三路遗址，灰坑）	10	-7.1 ± 1.2	8.5 ± 0.8	司艺：《2500 BC-1000 BC中原地区家畜饲养策略与先民肉食资源消费》，中国科学院大学博士学位论文，2013年。
狗（殷墟铁三路遗址，灰坑）	2	-10.1 ± 0.4	9.2 ± 0.3	
鼠（殷墟遗址，灰坑）	11	-9.2 ± 1.8	8.3 ± 0.5	Cheung C., Jing Z., Tang J., et al. 2017. Examining social and cultural differentiation in early Bronze Age China using stable isotope analysis and mortuary patterning of human remains at Xin'anzhuang, Yinxu. *Archaeological and Anthropological Sciences* 9 (5): 799-816.
人（殷墟遗址，墓葬）	32	-7.4 ± 0.8	/	张雪莲、王金霞、冼自强等：《古人类食物结构研究》，《考古》2003年第2期，第158~171页。
人（殷墟遗址，灰坑和地层）	6	-8.7 ± 1.2	/	
人（殷墟遗址，墓葬）	1	-20.7	/	
人（殷墟遗址，不详）	1	/	5.9	
人（殷墟孝民屯遗址，墓葬）	3	-11.9 ± 3.2‰	9.7 ± 0.7	司艺：《2500 BC-1000 BC中原地区家畜饲养策略与先民肉食资源消费》，中国科学院大学博士学位论文，2013年。
人（殷墟孝民屯遗址，灰坑）	1	-10.1	9.0	

续表 2

骨骼的种属及出土背景	个体数	$\delta^{13}C \pm SD$ (‰)	$\delta^{15}N \pm SD$ (‰)	数据来源
人（殷墟遗址及同时临近遗址，墓葬）	55	-9.0 ± 1.1	9.9 ± 0.8	a. Cheung C., Jing Z., Tang J., et al. 2017. Examining social and cultural differentiation in early Bronze Age China using stable isotope analysis and mortuary patterning of human remains at Xin'anzhuang, Yinxu. *Archaeological and Anthropological Sciences* 9 (5): 799–816. b. Cheung C., Jing Z., Tang J., et al. 2017. Diets, social roles, and geographical origins of sacrificial victims at the royal cemetery at Yinxu, Shang China: new evidence from stable carbon, nitrogen, and sulfur isotope analysis. *Journal of Anthropological Archaeology* 48: 28–45. c. Cheung C., Jing Z., Tang J., et al. 2017. Social dynamics in Early Bronze Age China: a multi-isotope approach. *Journal of Archaeological Science: Reports* 16: 90–101.
人（殷墟遗址人牲和人殉，墓葬）	64	-7.9 ± 0.5	8.2 ± 0.7	Cheung C., Jing Z., Tang J., et al. 2017. Diets, social roles, and geographical origins of sacrificial victims at the royal cemetery at Yinxu, Shang China: new evidence from stable carbon, nitrogen, and sulfur isotope analysis. *Journal of Anthropological Archaeology* 48: 28–45.

经过对殷墟遗址及其周边遗址同时期家猪、家犬的 C、N 稳定同位素值的梳理和比较，可以发现这一时期先民和家猪、家犬的食物结构依然存在较强的相关性，即表现出较强依赖粟黍农业的传统。然而随着社会复杂化进程的加剧及农作物种类的多样化发展，不同人群的食物结构呈现出一定的差异，将家猪和家犬的食物结构作为先民食物结构的替代性指标可能会存在较大的误差。反观这一时期鼠和部分先民的 C、N 稳定同位素值存在着更强的相关性，但是有关鼠的研究在我国较少，没有足够的案例支撑更为深入的讨论。

鉴于人和家猪、家犬在营养级上的固有差异，他们在 N 稳定同位素值上存在约一个营养级的差异（约富集 3‰~5‰），因此相关探讨存在一定的不确定性。而 C 稳定同位素值在营养级间传递过程中发生的分馏效应（约富集 1‰~1.5‰）常常可以忽略不计[35]，因此对 C 稳定同位素的相关值进行讨论可以较好反映人和家猪、家犬食物来源（包括植物或动物）类型的相关性。

为了更清楚地还原人和家猪、家犬食物来源（包括植物或动物）类型的相关性，我们对辉县孙村遗址家猪、家犬以及安阳殷墟遗址的先民、家猪、家犬的 C 稳定同位素的均值做棒状误差图（图2）。需要指出的是，表2 中出现了一个特殊值（−20.7‰）和一个埋藏单位特殊的个体（灰坑），暂不对其进行统计。

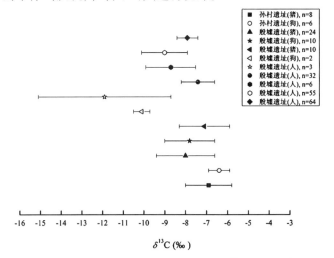

图2　孙村遗址及殷墟遗址人、家猪、家犬的 δ^{13}C 平均值棒状误差图
Figure 2. Error bar of δ^{13}C of humans, pigs and dogs from Suncun and Yinxu.

如图2所示，辉县孙村遗址家猪和家犬的 δ^{13}C 均值总体上略高于同一时期殷墟遗址的家猪、家犬和先民，说明辉县孙村遗址的家猪和家犬更加依赖粟黍和/或副产品，而殷墟遗址家猪、家犬和先民的食物结构中则明显有 C_3 类食物。同时，殷墟遗址个别先民和家犬的 δ^{13}C 均值明显低于大多数先民、家犬和家猪的相应值，说明他们的食物结构中 C_3 类食物比重较大。此外，殷墟遗址不同人群的 δ^{13}C 均值也存在一定的差异，如同一墓地人牲和人殉的 δ^{13}C 均值要明显高于墓主，再如不同出土单位人群的 δ^{13}C 均值也存在差异，说明殷墟遗址不同人群食物结构的差异已经逐渐明晰。

尽管殷墟遗址先民与家猪、家犬的食物结构依然存在较强的相关性，但不同人群的食物结构已呈现出明显的差异，因此家猪、家犬的食物结构不能反映先民日益复杂化的生业变化。因此，拟用辉县孙村遗址家猪和家犬的食物结构来还原当地先民的生业经济可能存在一定的困难，但可以初步推测当地的生业经济以粟黍为主。

（五）家猪和家犬在重建先民食物结构中的指示性作用

新石器时代中晚期，中国北方地区家猪、家犬及先民的食物结构表现出极强的相关性，特别是食物来源（包括植物或动物）的类型基本一致。这种极强的相关性主要是由于粟黍农业的发展、强化和扩张，先民及其饲喂的家猪、家犬越来越单一的依赖粟黍及其副产品。因此，家猪和家犬在重建先民食物结构中具有重要的指示性作用，可以作为重建先民食物结构的替代性指标[36]。

龙山文化晚期及以后，北方地区粟黍农业的地位逐步受到外来农作物和其他新的农作物的冲击，先民的食物结构也开始发生相应的变化[37]。如上文分析所示，不论殷墟遗址先民的食物结构，还是殷墟遗址家猪和家犬的食物结构，都明显添加了部分 C_3 类食物，但他们的食物结构依然存在一定的相关性。也就是说，家猪和家犬的食物结构依然可以作为大部分群体食物结构的一种间接、笼统的指示参考，但不能完全作为重建先民食物结构的替代性指标。因此，辉县孙村遗址家猪和家犬的食物结构可以为当地先民的食物结构和生业经济的重建提供一个重要参考。

值得注意的是，部分先民的食物结构存在明显的差异，特别是不同年龄、性别和阶层人群的食物结构可能存在较大差异[38]，如表2所列数据中，人牲、人殉的食物结构明显不同于其他先民。而几乎每一个遗址总有一些特殊的个体，如表2中某个个体的 $\delta^{13}C$ 值为 -20.7‰。显然，将家猪和家犬的食物结构作为指示先民食物结构的指标并不能涵盖这些特殊信息。此外，在越来越复杂的社会进程中，家鼠与大多数农耕先民的食物结构相关性更强，也是值得关注和研究的问题之一。

四 结论

通过对孙村遗址殷商文化时期动物骨的 C、N 稳定同位素分析，我们重建了动物的食物结构，探讨了当时的家畜饲养情况，同时检验并审视了家猪和家犬是否可以继续作为重建先民生业经济的替代性指标。结论如下：

孙村遗址猪、狗、牛的食物主要为 C_4 类食物，即粟黍及其副产品等，说明主要依赖粟黍农业饲喂；羊的数据较少，但食物结构差异较大，说明先民可能在野外放养羊，也可能依赖粟黍农业喂养羊；鹿主要以 C_3 类野生草本植物为食，说明鹿主要在野外生活，应该和先民的管理无关。

比较孙村遗址及其临近的殷墟遗址先民、家猪、家犬的稳定同位素值，结果显示家猪和家犬的食物结构可以作为殷商文化时期大部分先民食物结构的一种间接、笼统的指示参考。

随着社会复杂化程度的加剧，外来及新的农作物的推广和种植，以及家畜专业化饲喂和管理的强化，家猪和家犬与先民的食物结构相关性变得越来越复杂，将家猪和家犬的食物结构作为重建先民生业经济参考时需要更加谨慎。

注释

[1] Yang X., Wan Z., Perry L., et al. 2012. Early millet use in northern China. *Proceedings of the National Academy of Sciences of the United States of America* 109（10）：3726 – 3730.

[2] 袁靖：《中国古代家养动物的动物考古学研究》，《第四纪研究》2010 年第 30 卷第 2 期，第 298～306 页。

[3] 董广辉、张山佳、杨谊时等：《中国北方新石器时代农业强化及对环境的影响》，《科学通报》2016 年第 61 卷第 26 期，第 2913～2925 页。

[4] 侯亮亮：《稳定同位素视角下重建先民生业经济的替代性指标》，《南方文物》2019 年第 2 期，第 165～183 页。

[5] 侯亮亮：《稳定同位素视角下重建先民生业经济的替代性指标》，《南方文物》2019年第2期，第165~183页。

[6] 侯亮亮：《稳定同位素视角下重建先民生业经济的替代性指标》，《南方文物》2019年第2期，第165~183页。

[7] Kohn M. J. 1999. You are what you eat. *Science* 283 (5400): 335–336.

[8] Hu Y. 2018. Thirty-four years of stable isotopic analyses of ancient skeletons in China: An overview, progress and prospects. *Archaeometry* 60 (1): 144–156.

[9] 侯亮亮：《稳定同位素视角下重建先民生业经济的替代性指标》，《南方文物》2019年第2期，第165~183页。

[10] 侯亮亮：《稳定同位素视角下重建先民生业经济的替代性指标》，《南方文物》2019年第2期，第165~183页。

[11] a. 中国社会科学院考古研究所：《中国考古学·夏商卷》，中国社会科学出版社，2003年。
b. 井中伟、王立新：《夏商周考古学》，科学出版社，2013年。
c. 游修龄：《中国农业通史·原始社会卷》，中国农业出版社，2008年。
d. 陈文华：《中国农业通史·夏商西周春秋卷》，中国农业出版社，2007年。

[12] a. 赵志军：《小麦传入中国的研究——植物考古资料》，《南方文物》2015年第3期，第44~52页。
b. Liu X., Lister D. L., Zhao Z., et al. 2017. Journey to the east: Diverse routes and variable flowering times for wheat and barley en route to prehistoric China. *PLOS ONE* 12 (11): e0187405.
c. 陈文华：《中国农业通史·夏商西周春秋卷》，中国农业出版社，2007年。

[13] 陈文华：《中国农业通史·夏商西周春秋卷》，中国农业出版社，2007年。

[14] 袁靖：《生业兴起：文明进程中的五谷、六畜、百工》，《中国文化遗产》2012年第4期，第40~46页。

[15] a. Dong Y., Morgan C., Chinenov Y., et al. 2017. Shifting diets and the rise of male-biased inequality on the Central Plains of China during Eastern Zhou. *Proceedings of the National Academy of Sciences of the United States of America* 114 (5): 932–937.
b. Liu X., Lightfoot E., O'Connell T. C., et al. 2014. From necessity to choice: Dietary revolutions in west China in the second millennium BC. *World Archaeology* 46 (5): 661–680.

[16] 郑州大学历史学院考古系、河南省文物管理局"南水北调"办公室、新乡市文物局等：《河南辉县孙村遗址发掘简报》，《中原文物》2008年第1期，第4~18页。

[17] 河南省文物局：《辉县孙村遗址》，科学出版社，2012年。

[18] Richards M. P., Hedges R. E. M. 1999. Stable isotope evidence for similarities in the types of marine foods used by Late Mesolithic humans at sites along the Atlantic coast of Europe. *Journal of Archaeological Science* 26 (6): 717–722.

[19] Hedges R. E. M., 2002. Bone diagenesis: An overview of processes. *Archaeometry* 44 (3): 319–328.

[20] Ambrose S. H., Butler B. M., Hanson D. H., et al., 1997. Stable isotopic analysis of human diet in the Marianas Archipelago, Western Pacific. *American Journal of Physical Anthropology* 104 (3): 343–361.

[21] Ambrose S. H. 1990. Preparation and characterization of bone and tooth collagen for isotopic analysis. *Journal of Archaeological Science* 17 (4): 431–451.

[22] DeNiro M. J. 1985. Post-mortem preservation of alteration of in vivo bone collagen isotope ratios in relation to palaeodietary reconstruction. *Nature* 317 (6040): 806 – 809.

[23] Ambrose S. H. 1991. Effects of diet, climate and physiology on nitrogen isotope abundances in terrestrial foodwebs. *Journal of Archaeological Science* 18 (3): 293 – 317.

[24] Makarewicz C. A. 2015. Sealy J. Dietary reconstruction, mobility, and the analysis of ancient skeletal tissues: Expanding the prospects of stable isotope research in archaeology. *Journal of Archaeological Science* 56: 146 – 158.

[25] a. 中国社会科学院考古研究所:《中国考古学·夏商卷》,中国社会科学出版社,2003 年。
b. 井中伟、王立新:《夏商周考古学》,科学出版社,2013 年。
c. 游修龄:《中国农业通史·原始社会卷》,中国农业出版社,2008 年。
d. 陈文华:《中国农业通史·夏商西周春秋卷》,中国农业出版社,2007 年。

[26] 陈相龙、袁靖、胡耀武等:《陶寺遗址家畜饲养方式策略初探:来自碳、氮稳定同位素的证据》,《考古》2012 年第 9 期,第 75 ~ 82 页。

[27] 周振甫:《诗经译注》,中华书局,2002 年。

[28] 河南省文物局:《辉县孙村遗址》,科学出版社,2012 年。

[29] 张雪莲、王金霞、冼自强等:《古人类食物结构研究》,《考古》2003 年第 2 期,第 158 ~ 171 页。

[30] 司艺:《2500 BC – 1000 BC 中原地区家畜饲养策略与先民肉食资源消费》,中国科学院大学博士学位论文,2013 年。

[31] Cheung C., Jing Z., Tang J., et al. 2017. Examining social and cultural differentiation in early Bronze Age China using stable isotope analysis and mortuary patterning of human remains at Xin'anzhuang, Yinxu. *Archaeological and Anthropological Sciences* 9 (5): 799 – 816.

[32] Cheung C., Jing Z., Tang J., et al. 2017. Examining social and cultural differentiation in early Bronze Age China using stable isotope analysis and mortuary patterning of human remains at Xin'anzhuang, Yinxu. *Archaeological and Anthropological Sciences* 9 (5): 799 – 816.

[33] a. 闫灵通:《稳定同位素在陆生动物矿化组织研究中的应用》,中国科学院研究生院博士学位论文,2010 年。
b. 司艺:《2500 BC – 1000 BC 中原地区家畜饲养策略与先民肉食资源消费》,中国科学院大学博士学位论文,2013 年。

[34] Cheung C., Jing Z., Tang J., et al. 2017. Examining social and cultural differentiation in early Bronze Age China using stable isotope analysis and mortuary patterning of human remains at Xin'anzhuang, Yinxu. *Archaeological and Anthropological Sciences* 9 (5): 799 – 816.

[35] a. Ambrose S. H., Norr L. 1993. Isotopic composition of dietary protein and energy versus bone collagen and apatite: Purified diet growth experiments. In: Lambert J. B., Grupe G. (eds.), *Prehistoric Human Bone: Archaeology at the Molecular Level*. Berlin: Springer-Verlag, pp. 1 – 37.
b. Hedges R. E. M., Reynard L. M. 2007. Nitrogen isotopes and the trophic level of humans in archaeology. *Journal of Archaeological Science* 34: 1240 – 1251.

[36] 侯亮亮:《稳定同位素视角下重建先民生业经济的替代性指标》,《南方文物》2019 年第 2 期,第 165 ~ 183 页。

[37] a. 赵志军:《小麦传入中国的研究——植物考古资料》,《南方文物》2015 年第 3 期,第 44 ~ 52 页。

  b. Liu X., Lister D. L., Zhao Z., et al. 2017. Journey to the east: Diverse routes and variable flowering times for wheat and barley en route to prehistoric China. *PLOS ONE* 12（11）: e0187405.

  c. 陈文华:《中国农业通史·夏商西周春秋卷》,中国农业出版社,2007年。

[38] a. Cheung C., Jing Z., Tang J., et al. 2017. Examining social and cultural differentiation in early Bronze Age China using stable isotope analysis and mortuary patterning of human remains at Xin'anzhuang, Yinxu. *Archaeological and Anthropological Sciences* 9（5）: 799 – 816.

  b. Cheung C., Jing Z., Tang J., et al. 2017. Diets, social roles, and geographical origins of sacrificial victims at the royal cemetery at Yinxu, Shang China: New evidence from stable carbon, nitrogen, and sulfur isotope analysis. *Journal of Anthropological Archaeology* 48: 28 – 45.

  c. Cheung C., Jing Z., Tang J., et al. 2017. Social dynamics in Early Bronze Age China: A multi-isotope approach. *Journal of Archaeological Science: Reports* 16: 90 – 101.

黄牛与牦牛骨骼形态的对比观察*

Differences in Osteological Morphology between Cattle (*Bos taurus*) and Yak (*Bos grunniens*)

李 凡

Fan Li

郑州大学历史学院，郑州

School of History, Zhengzhou University, Zhengzhou, China

摘要：黄牛和牦牛的骨骼区分在青藏高原考古遗址的动物考古学研究中具有重要意义。本文主要通过观察和测量，对黄牛和牦牛的前颌骨、下颌骨、寰椎、枢椎、掌骨、跖骨、第一指/趾节骨、第二指/趾节骨等骨骼部位的形态差异及相关数据进行了对比。

关键词：黄牛；牦牛；骨骼形态；对比

Abstract: The distinguishing between the skeletal elements of cattle (*Bos taurus*) and yak (*Bos grunniens*) is of great significance for zooarchaeologial study of the Tibetan Plateau. Skeletal elements from at least nine individuals of cattle and yak were observed and measured, respectively (not all of them are complete skeletons, i.e., some elements are missing). Identifiable features between cattle and yak skeletons can be recognized on the premaxilla, mandible, atlas, axis, metacarpus, metatarsus, humerus, first and second phalanges.

1. The corner between the anterior and lateral edges of the premaxilla in cattle is rounded (Plate XXV. Figure 1. a & c), while in yak is sharp (Plate XXV. Figure 1. b & d).

2. The anterior part of the mandible in cattle (Plate XXV. Figure 2. a) shows an obviously larger up-warping angle than in yak (Plate XXV. Figure 2. b).

3. The dorsal tubercle of the atlas in cattle is more robust, with the cranial part being more pointed and prominent (Plate XXVI. Figure 1. a); the dorsal tubercle of the atlas in yak is smaller, with the cranial part being relatively blunt (Plate XXVI. Figure 1. b).

4. The overall shape of the axis in cattle is relatively narrow and long, and the caudal part of its ventral crest protrudes prominently and extends (Plate XXVI. Figure 2. a); the axis of yak is relatively wide and short, and the caudal part of its ventral vest is less prominent (Plate XXVI.

---

* 本研究由国家社会科学基金青年项目（18CKG013）和郑州中华之源与嵩山文明研究会青年课题（Q2017-5）资助。

Figure 2. b).

5. The nodule, which is located above the depression for the attachment of lateral ligament in the distal end of the humerus, protrudes proximodorsally clearly in the shape of "6" in cattle (Plate XXVII. Figure 1. a), while presents much weaker in yak (Plate XXVII. Figure 1. b).

6. The overall shape of the metacarpus in cattle is narrow and long (Plate XXVII. Figure 2. a), while in yak is wide and short (Plate XXVII. Figure 2. b). Besides, a notch is observed between the ridge of the base and the proximal metacarpal canal in cattle (Plate XXVIII. Figure 1. a & Figure 2. a), but is not present in yak (Plate XXVIII. Figure 1. b & Figure 2. b).

7. The overall shape of the metatarsus in cattle is narrow and long (Plate XXVIII. Figure 3. a), while in yak is wide and short (Plate XXVIII. Figure 3. b).

8. The overall shape of the first phalanx in cattle is narrow and long (Plate XXIX. Figure 1. a), while in yak is wide and short (Plate XXIX. Figure 1. b).

9. The overall shape of the second phalanx in cattle is narrow and long (Plate XXIX. Figure 2. a), while in yak is wide and short (Plate XXIX. Figure 2. b). Besides, the base is relatively narrower (Plate XXX. a), while in yak is wider (Plate XXX. b).

Keywords: cattle, yak, osteolgical morphology, comparative observation

黄牛是中国传统"六畜"之一，可适应多种生态环境，分布广泛，在国内的众多古代遗址中都能见到黄牛的骨骼遗存。家养黄牛由野牛驯化而来，至少在距今4500~4000年前后出现于黄河中下游地区，但在中国境内的起源时间和地点尚不明确[1]。

牦牛主要分布于高寒高山草原地区，在我国以青藏高原为主要分布地区。家养牦牛由野牦牛驯化而来，但驯化开始的时间和地点尚不明确，目前发现牦牛可能被人类利用的最早证据出自距今3700~3500年的拉萨曲贡遗址[2]。

黄牛和牦牛在动物界分类系统中的位置如下所示：

脊索动物门　Chordata
　哺乳纲　Mammalia
　　偶蹄目　Artiodactyla
　　　牛科　Bovidae
　　　　牛属　*Bos*
　　　　　黄牛　*Bos taurus* Linnaeus, 1758
　　　　　牦牛　*Bos grunniens* Linnaeus, 1766

黄牛和牦牛是牛属（*Bos*）中不同种的动物，二者骨骼的构成和基本形态大体相同，但某些骨骼存在可观察到的形态（包括形状、三维比例等）差异。在青藏高原考古遗址出土动物遗存的考古学研究中，黄牛和牦牛的骨骼鉴别具有重要意义。但迄今为止，我国的动物考古学研究中尚未有关于黄牛和牦牛骨骼区别的研究成果正式发表。

笔者参考了河南省文物考古研究院动物考古实验室和中国社会科学院考古研究所科技考古中心动物考古实验室收藏的黄牛和牦牛现生骨骼标本，对肉眼所能观察到的二者骨骼形态区别进行一一比对，其中黄牛标本至少16个个体，牦牛标本至少10个个体，但并非全部为完整个体，即有些个体存在骨骼缺失。文中所涉骨骼部位及方位多按照动物解剖学通用术语，具体可参看《考古遗址出土动物骨骼测量指南》[3]，文中所涉及测量点也以此书为依据。

一　前颌骨

前颌骨在上颌骨的前端。黄牛前颌骨前端与外侧之间的转角呈斜抹状，几乎无棱角（图版25：图1，a、c）。牦牛前颌骨前端与外侧之间的转角近似方形，棱角明显（图版25：图1，b、d）。且无论从背侧或腹侧观均可看出二者区别。

二　下颌骨

黄牛和牦牛下颌骨的区别主要表现在其前部上翘角度有所差异。将下颌水平放置进行观察和测量，由于此类测量并不能非常精确，故只将数值保留至个位。黄牛下颌骨前部上翘角度较大（图版25：图2，a），测量8件标本，上翘角度在21°～36°，平均上翘26°。牦牛下颌骨前部上翘角度较小、较平缓（图版25：图2，b），测量9件标本，上翘角度在6°～16°，平均上翘11°（表1）。

表1　黄牛和牦牛下颌骨前端上翘角度测量值
Table 1　Up-warping angle of anterior part of cattle and yak mandibles.

黄牛 Cattle		牦牛 Yak	
标本编号 No.	角度 Angle	标本编号 No.	角度 Angle
黄58	26°	牦291	12°
黄61	22°	牦292	11°
黄60	26°	牦293	12°
黄78	24°	牦294	10°
黄93	21°	牦296	15°～16°
黄230	30°	牦298	10°
黄75	25°	牦299	6°
黄92	36°	牦300	12°
		牦301	13°
平均 Average	26°	平均 Average	11°

三　寰椎

寰椎是颅后的第一节颈椎，其前端与枕骨髁相接合，背侧有骨质凸起，称为背侧结节。黄牛的寰椎背侧结节较大，其前端较尖，明显凸起（图版26：图1，a）。牦牛的背侧结节相对较小，前端较钝（图版26：图1，b）。

四　枢椎

枢椎是第二节颈椎，其前端与寰椎后端相接合，后接第三节颈椎。黄牛的枢椎整体较窄长，腹侧嵴的尾端凸出明显，向后延伸较多（图版26：图2，a）。牦牛整体较宽短，腹侧嵴尾端凸出程度较弱（图版26：图2，b）。

五　肱骨

肱骨远端外侧髁的凹陷上方有一凸起的结节。黄牛的这一结节向前上方明显呈略尖状凸出，与凹陷形成状如"6"的形态（图版27：图1，a）。牦牛的结节较不明显（图版27：图1，b）。

六　掌骨

从整体形态上看，黄牛掌骨较窄长（图版27：图2，a），测量9件成年或亚成年标本，长度（GL）最小值为200.17 mm，最大值为233.41 mm，平均值为215.84 mm；长宽比（GL/SD）最小值为5.32，最大值为8.38，平均值为6.82。而牦牛掌骨整体较宽短（图版27：图2，b），测量6件成年或亚成年标本，长度（GL）最小值为143.19 mm，最大值为153.81 mm，平均值为149.07 mm；长宽比（GL/SD）最小值为3.62，最大值为5.03，平均值为4.30。而二者的最小宽（SD）数值范围交集较多，参考价值不大，因此主要以长度和长宽比作为观察依据（表2）。

除掌骨整体形态差异之外，黄牛和牦牛掌骨近端形态也存在区别。黄牛掌骨近端隆起与血管沟相接处有一道沟（图版28：图1，a），从近端关节面看，前侧中间部位有一凹陷（图版28：图2，a）。而牦牛掌骨近端隆起与血管沟相接处无沟（图版28：图1，b），从近端关节面看，前侧中间部位无凹陷（图版28：图2，b）。

七　跖骨

从整体形态上看，黄牛跖骨较窄长（图版28：图3，a），测量9件成年或亚成年标本，长度（GL）最小值为224.27 mm，最大值为263.62 mm，平均值为246.06 mm；长宽比（GL/SD）最小值为7.18，最大值为10.25，平均值为9.02。而牦牛跖骨整体较宽短（图版 28：图 3，b），测量 6 件成年或亚成年标本，长度（GL）最小值为 179.77 mm，最大值为 188.47 mm，平均值为 185.41 mm；宽比（GL/SD）最小值为 5.96，最大值为 8.02，平均值为 6.88。而二者的最小宽（SD）数值范围交集较多，参考价值不大，因此主要以长度和长宽比作为观察依据（表 3）。

表 2 黄牛与牦牛掌骨长宽测量值
Table 2. Length (GL) and the smallest breath (SD) of large metacarpus of cattle and yak (in mm).

黄牛掌骨 Cattle				牦牛掌骨 Yak			
标本编号 No.	长 GL (mm)	最小宽 SD (mm)	长宽比 GL/SD	标本编号 No.	长 GL (mm)	最小宽 SD (mm)	长宽比 GL/SD
黄 82	230.27	38.55	5.97	牦 292	145.82	28.97	5.03
黄 78	200.17	23.88	8.38	牦 294	143.19	31.83	4.50
黄 83	219.72	32.50	6.76	牦 295	153.81	34.33	4.48
黄 75	214.95	31.99	6.72	牦 297	149.48	37.64	3.97
黄 84	233.41	36.96	6.32	牦 293	151.02	41.75	3.62
黄 94	210.48	29.77	7.07	牦 302	151.07	35.96	4.20
黄 93	209.87	29.34	7.15				
黄 72	210.69	39.58	5.32				
黄 69	213.01	27.85	7.65				
平均 Average	215.84	32.27	6.82	平均 Average	149.07	35.08	4.30

表 3 黄牛与牦牛跖骨长宽测量值
Table 3. Lengths (GL) and the smallest breath (SD) of large metatarsus of cattle and yak.

黄牛跖骨 Cattle				牦牛跖骨 Yak			
标本编号 No.	长 GL (mm)	最小宽 SD (mm)	长宽比 GL/SD	标本编号 No.	长 GL (mm)	最小宽 SD (mm)	长宽比 GL/SD
黄 78	224.27	21.89	10.25	牦 292	182.32	22.72	8.02
黄 82	257.60	35.88	7.18	牦 297	187.34	30.14	6.22
黄 83	247.03	29.32	8.43	牦 293	187.70	31.50	5.96
黄 84	263.62	31.80	8.29	牦 295	188.47	27.38	6.88
黄 93	247.04	25.62	9.64	牦 302	186.86	27.10	6.90
黄 92	243.06	23.77	10.23	牦 294	179.77	24.74	7.27
黄 231	242.83	28.08	8.65				
黄 75	244.40	27.72	8.82				
黄 60	244.70	25.24	9.69				
平均 Average	246.06	27.70	9.02	平均 Average	185.41	27.26	6.88

八 第一指/趾节骨

从整体形态上看，黄牛的第一指/趾节骨较窄长（图版29：图1，a），测量7件分别来自7个成年或亚成年个体的标本，长度（GLpe）最小值为54.86 mm，最大值为71.97 mm，平均值为64.50 mm；长宽比（GLpe/SD）最小为2.36，最大为2.82，平均值为2.60。而牦牛的第一指/趾节骨整体较宽短（图版29：图1，b），测量5件分别来自5个成年或亚成年个体的标本，长度（GLpe）最小值为49.79 mm，最大值为54.37 mm，平均值为52.28 mm；长宽比（GLpe/SD）最小为1.77，最大为2.38，平均值为2.04。而二者的最小宽（SD）数值范围交集较多，参考价值不大，因此主要以长度和长宽比作为观察依据（表4）。

表4 黄牛与牦牛第一指/趾节骨长宽测量值
Table 4. Length (GLpe) and the smallest breath (SD) of first phalanx of cattle and yak.

黄牛第一指/趾节骨 Cattle				牦牛第一指/趾节骨 Yak			
标本编号 No.	长 GLpe (mm)	最小宽 SD (mm)	长宽比 GLpe/SD	标本编号 No.	长 GLpe (mm)	最小宽 SD (mm)	长宽比 GLpe/SD
黄75	62.33	24.11	2.59	牦292	49.79	24.82	2.01
黄78	54.86	20.00	2.74	牦293	53.21	29.99	1.77
黄61	68.27	26.69	2.56	牦294	51.11	27.15	1.88
黄84	71.61	26.81	2.67	牦295	54.37	22.85	2.38
黄231	66.57	23.59	2.82	牦302	52.90	24.29	2.18
黄232	55.92	22.66	2.47				
黄82	71.97	30.49	2.36				
平均 Average	64.50	24.91	2.60	平均 Average	52.28	25.82	2.04

九 第二指/趾节骨

从整体形态上看，黄牛第二指/趾节骨较窄长（图版29：图2，a），测量7件成年或亚成年标本，长度（GL）最小值为38.81 mm，最大值为47.87 mm，平均值为44.33 mm；长宽比（GL/SD）最小为1.72，最大为2.15，平均值为1.96。牦牛第二指/趾节骨整体较宽短（图版29：图2，b），测量5件成年或亚成年标本，长度（GL）最小值为33.21 mm，最大值为38.74 mm，平均值为35.21 mm；长宽比（GL/SD）最小为1.43，最大为1.57，平均值为1.49。而二者的最小宽（SD）数值范围交集较多，参考价值不大，因此主要以长度和长宽比作为观察依据（表5）。

此外，黄牛和牦牛第二指/趾节骨近端关节面的形态也有差异。黄牛的近端关节面最大宽（BFp）较窄（图版30：a），而牦牛较宽（图版30：b）。但测量数据并不能很理想地表现这一点。

表5 黄牛与牦牛第二指/趾节骨长宽测量值
Table 5. Length (GLpe) and the smallest breath (SD) of second phalanx of cattle and yak.

黄牛第二指/趾节骨 Cattle				牦牛第二指/趾节骨 Yak			
标本编号 No.	长 GL (mm)	最小宽 SD (mm)	长宽比 GL/SD	标本编号 No.	长 GL (mm)	最小宽 SD (mm)	长宽比 GL/SD
黄75	43.69	22.05	1.98				
黄78	39.05	18.13	2.15	牦292	33.21	21.19	1.57
黄61	46.03	23.79	1.93	牦293	38.74	25.59	1.51
黄84	47.87	24.30	1.97	牦294	34.13	23.10	1.48
黄231	47.54	22.31	2.13	牦295	34.79	24.33	1.43
黄232	38.81	21.38	1.82	牦302	35.18	24.51	1.44
黄82	47.32	27.59	1.72				
平均 Average	44.33	22.79	1.96	平均 Average	35.21	23.74	1.49

以上所列黄牛和牦牛的骨骼形态差异，是基于数量有限的现生黄牛和牦牛标本比对得出的初步认识。笔者曾对西藏阿里地区故如甲木墓地和曲踏墓地出土的动物骨骼进行鉴定，其中出土的黄牛和牦牛骨骼并未包括以上所有部位。根据前文及图、表所示，黄牛和牦牛多数骨骼部位的区别较为明显，在鉴定实践中较容易把握。笔者仅对故如甲木墓地中鉴定为牦牛的2件寰椎及鉴定为黄牛的1件寰椎进行了DNA实验分析，实验结果与鉴定结果相符。因此，笔者认为以上方法可用于今后黄牛和牦牛骨骼鉴定的实践之中，并接受进一步检验。囿于笔者水平和观察标本数量，难免缺陷甚或错误，望广大同仁批评指正，并期待动物考古学界今后在同一问题上能有更多观察及收获。

致谢：本文所涉及比对标本主要来自河南省文物考古研究院动物考古实验室及中国社会科学院考古研究所动物考古实验室，文中所用图片的拍摄和后期处理均由笔者完成。古DNA分析由中国社会科学院考古研究所DNA实验室的赵欣博士完成。在此向提供帮助的各位同仁特致谢忱！

注释

[1] 袁靖：《中国古代家养动物的动物考古学研究》，《第四纪研究》2010年第2期，第298～306页。

[2] 周本雄：《曲贡遗址的动物遗存》，中国社会科学院考古研究所、西藏自治区文物局编著《拉萨曲贡》，中国大百科全书出版社，1999年，第237～243页。

[3] 安格拉·冯登德里施著，马萧林、侯彦峰译：《考古遗址出土动物骨骼测量指南》，科学出版社，2007年。

考古遗址出土啮齿目遗存的采集与鉴定方法
Methods of Collection and Identification of Rodent Remains from Archaeological Sites

王运辅

Yunfu Wang

重庆师范大学科技考古实验室，重庆

Laboratory of Scientific Archaeology, Chongqing Normal University, Chongqing, China

摘要：考古遗址出土啮齿目遗存对于讨论家鼠与人共栖起源与演变、认识古代人类生活及与动物关系等具有重要的意义，相关的采集与鉴定工作构成了深入研究的基础。本文首先介绍了在考古遗址中采集啮齿目遗存的基本技术，然后以褐家鼠、黑家鼠、黄胸鼠、小家鼠等家栖鼠类，以及多种与人类生活密切相关的野生啮齿动物为实例，讨论了啮齿目鉴定工作中应该重点观察的若干解剖特征点，包括头骨的基本形态、各种脊、突、孔、窝等细微结构，以及基本的颊齿类型等。本文所归纳的采集与鉴定方法简明扼要，适用于动物考古学研究的实际需要。

关键词：啮齿目；鉴定；头骨形态；脊、突、孔、窝

Abstract: Remains of rodents from archaeological sites are of central significance for understanding the origin and evolution of commensalism between human and house rodents, as well as the dynamics of ancient human lifeways and human-animal relationships. Consequently, the methods of collection and identification are an important focus for research and have immense practical value in zooarchaeological fieldwork. In this article, I discuss methods for collection and identification of rodent skeletal materials. The remains of wild and commensal rodents are used to discuss a series of distinguishing anatomical points, including skull shape, micro structures such as crests, processes, foramens and fovea, and basic types of molars.

Rodent remains are normally present at Chinese archaeological sites in several forms: 1) Incisors and cheek teeth, separated from the skulls-the most common type of discovery; 2) Mandibles, usually with some incisors or lower cheek teeth; 3) Broken parts of skulls, especially hard palates containing some upper cheek teeth; 4) Some solid post-cranial bones such as femurs. All such samples are valuable in the identification of rodents. High potential localities for recovery of rodent bones should be carefully inspected during excavations-these include food cellars, ash-pits and drainage passages. Furthermore, soil containing rodent remains can be transported to the laboratory for thorough and fine sorting.

Some key points can be followed when observing the skulls of rodents. First of these is the morphology of the skull. Generally, a bulbous skull may belong to a member of the squirrel family, while a low and flat skull represents burrowing mice or rats. Next, the shape of the zygomatic arch is also informative. For instance, three different bamboo rats found in South China can be differentiated by referring to their zygomatic arches. Next, the length and width of snout bones often reveal something about the animals' particular living habits.

Micro anatomical structures of skulls should also be a focus for rodent identification.

1) The appearance or absence, development or underdevelopment of the supraorbital crest is useful for distinguishing some similar rodents. For example, the skull of *Apodemus speciosus* is very similar with that of *Apodemus sylvaticus*—but the former presents a supraorbital crest while the latter does not. Similarly, the skull of *Rattus bowersi* shares many traits with that of *Leopoldamysedwardsi*. However, the supraorbital crest of the former is some thinner than that of the latter.

2) The lambdoidal and sagittal crests can also be valuable when comparing some species of Sciuridae, Rhizomvinae or Myospalacinae.

3) Both rough classifications and exact identification can be made on the basis of the angular process, condyloid process and coronoid process of the mandible. The coronoid processes of Sciuromorpha and Myomorpha rodents resemble each other. However, the condyloid process of Myomorpha rodents is much rounder and more blunt in shape, while that of Sciuromorpha rodents appears square with a small tubercle. The angular process of Sciuromorpha rodents also makes a forward extension forming a so-called subangular process, while that of Myomorpha rodents shows a weak protrusion. The angular process of the bamboo rat is similar to that of hedgehog, but can be easily distinguished from its much smaller size.

4) An incisor socket process is only found among some species of Murinae and Cricetinae, and is especially well-developed among all species of Rhizomvinae. Some other species of Murinae and Cricetinae instead develop weak bulges to accommodate their incisors. An example is that *Rattus norvegicus* has a well-developed incisor socket process, while some similar rats or mice such as *Rattus rattus*, *Rattus flavipectus*, and *Mus musculus* do not forms incisor socket processes.

5) Supraorbital process and preorbital process structures are only observed among Sciuridae species. According to the shape differences of these processes, some species of Sciuridae can be determined.

6) Postbital process on temporal bone are a useful marker to seek out some species of Arvicolinae and *Marmota*. For example, the grey marmot bears this process while other

marmots like Himalayan marmots do not.

7) The antorbital foramen is also significant for rodent identification. The antorbital foramen of Sciuromorpha rodents is very small, under which a process is formed to attach the masseter muscle. In contrast, the antorbital foramen of Myomorpha rodent is relatively larger, and that of Hystricomorpha tends to be nearly as large as its orbital. The antorbital foramen of Myomorpha rodents can be grouped into three types. These include the "key hole" type, a gap with relatively broad upper part and smaller lower part, resembling a key hole, which is present among species of Murinae. Next is the vertical triangular type, which is a typical trait among species of Myospalacinae. This type resembles a triangle with its long side vertically placed. Thirdly is the horizontal triangular type, which is unique to Rhizomvinae. This foramen appears like a triangle with its long side horizontally placed.

8) Snout bones showing Stensen's foramen, being useful for identification and worthy of careful study, are also common in archaeological sites. The length and width of Stensen's foramen, as well as the spacing between it and the front end of cheek teeth can be measured to distinguish different rodents. As an example, although the skull of *Apodemus speciosus* possesses many similarites with *Apodemus sylvaticus*, the spacing between Stensen's foramen and its first cheek tooth is clearly broader in *A. speciosus* than in *A. sylvaticus*.

9) The fovea posterior palatini is present among most species of Arvicolinae, which is an important marker to identify these rodents.

10) Finally, the hard palate fovea only appears on the *spermophilus*, which can be used to distinguish these squirrel-like rodents.

Incisors are the most common remains for rodents. Their shapes, structures, angles and colors can provide useful identification information. For instance, although it could not be possible by size and overall shape to distinguish the incisors of *Mus musculus* and *Apodemus sylvaticus*, the presence of a small step-like notch, which only exists on *M. musculus*, makes this separation possible.

Some post-cranial skeletons of rodents have been collected among some Chinese archaeological sites, however, we still lack useful comparative specimens and data for postcranial identifications. Another key logistic challenge in the identification of rodents is separating them from animals of similar size such as Insectivora and Chiroptera. Improving reference materials and comparative data in this regard will be an important focus of future work.

Keywords: Rodentia, identification, skull shape, crest, process, foramen and fovea

一　引言

对考古遗址出土的啮齿目遗存进行动物考古学研究，探讨家鼠与人共栖的起源、演变及由此伴生的鼠害等问题，是认识古代人类生活及与动物关系的一个重要方面，应该引起我们的高度重视。

做好考古遗址出土的啮齿目遗存的采集和鉴定工作是深入研究的基础，但目前尚存在一些难点：首先，由于啮齿目遗存材料细小，如果不对特定的出土单位进行水洗筛选，不易发现和收集，加之这类细小的动物遗存中还包含与啮齿目动物体型相似的食虫目、翼手目等动物，当它们混杂在一起时，没有专门的鉴定能力也难以区分。其次，国内专门研究啮齿目等小哺乳动物的研究人员不多，他们一般也不太关注考古遗址出土的此类动物遗存，而一般从事动物考古学研究的人员尚不具备对此进行鉴定的能力。其三，现有分类鉴定的参考文献不能满足动物考古学研究人员的需要。尽管国内多个省份都出版了当地啮齿目的动物志，汇总了大量的测量数据和野外实地观察报告，对于相关研究工作有重要的参考价值，但其图片幅面小，有些绘制粗糙，许多文字描述的细节未能在图片上表现清楚，对动物考古学研究人员而言在理解上有一定的难度。

有鉴于此，本文选取在考古遗址中最常见的褐家鼠、黑家鼠、黄胸鼠、小家鼠这4种家鼠，以及与人类活动有密切联系的松鼠、岩松鼠、黄鼠、旱獭、鼯鼠、仓鼠、䶄鼠、田鼠、沙鼠、竹鼠这10种野生啮齿目的现生标本进行观察，结合各类啮齿目动物志等参考文献，通过归纳和凝练，对较为常见的一些啮齿目动物的鉴定方法进行介绍。本文力求做到图片清晰、文字简明，用图文并茂的方式阐述上述啮齿目的一般结构特征，表现它们的主要解剖特征点，便于读者理解啮齿目头骨、牙齿及头后主要骨骼的基本结构，掌握鉴定啮齿目的主要观察点，为做好考古遗址出土啮齿目材料的鉴定工作奠定基础。

二　标本的采集

国内考古遗址大致已出土了32种啮齿目动物[1]，既有褐家鼠、黑家鼠、黄胸鼠、小家鼠等家鼠类，也有松鼠、岩松鼠、黄鼠、旱獭、鼯鼠、仓鼠、䶄鼠、田鼠、沙鼠、竹鼠等多种野生啮齿目，它们可能因人类的肉食或其他活动被带入遗址。今后的田野工作应当特别注意采集与人共栖的家鼠材料。

（一）啮齿目遗存的常见骨骼部位及保存状况

骨骼部位及保存状况与埋藏学过程有紧密的联系，可用于解读埋藏学成因[2]，例如小型猫科类动物在进食鼠类时常常弃置头部；猫头鹰等猛禽则习惯囫囵吞枣，其粪便堆积常包含完整的肢骨；而共栖在人类居址的鼠类在隐藏地点死亡后的堆积常包含鼠龄不一、较完整的骨骼组合。

结合实际发掘经验，啮齿目材料的常见出土部位及保存状况如下：

第一，下颌骨在遗址中的出现频率较高、多保存较好，常附着门齿和完整的臼齿齿列。此类材料特征明显，容易引起发掘者的注意。

第二，较完整的颅骨并不多见。颅骨遗存常仅有硬腭残段（硬腭是指前颌骨、上颌骨口腔一侧的骨片所构成的口腔骨板），常附着臼齿。而颅骨的纤薄脆弱部分，如顶骨、额骨以及颧弓等一般都残缺不见。

第三，头后骨骼以髋骨、肱骨、桡骨、尺骨、股骨、胫骨及腓骨最为常见，偶见椎骨。

第四，游离的门齿和臼齿常有较大的数量。

在采集时应该特别注意最具鉴定价值的颅骨、完整齿列等材料，也要注意采集游离的牙齿以及头后骨骼等，并注意与出土颅骨是否有明确的关联。

（二）采集技术

应结合啮齿目动物及其天敌的生活习性，注意观察考古遗址中可能出现啮齿目材料的地点。例如遗址中曾贮藏粮食的窖穴、弃置垃圾的灰坑、水流较缓的排水沟等等，都可能包含啮齿目材料。当发现啮齿目遗骸时，不要着急在现场刮取，应当按层位采集填土，分小袋包装，回到室内先拣选再水洗过筛进行收集。

三 鉴定准备与原则

鉴定准备工作主要有以下几点：1）清理标本；2）准备显微器材，手持放大镜以10倍、自带LED光源的鉴定镜为佳，另外准备放大倍率30倍以内带摄像头的体视显微镜；3）准备啮齿目分类文献，既要有总论性质的文献，也要有地方类的专题文献。

鉴定原则是依据清晰、测量与描述完整，只鉴定到分类依据可靠的分类层次。一般而言，只有带完整齿列且带若干可靠鉴定部位的颅骨类标本才可以鉴定到种，部分带完整齿列的下颌标本也可根据具体情况鉴定到种。游离的牙齿，尤其是头后骨骼，若不能明确与已知颅骨类标本的附属关系，依据国内分类学的基础条件，不宜鉴定到种；若要鉴定到种，需要对情况加以说明并做详细的测量与描述，以备其他研究者参考。

四 啮齿目的基本特征与分类地位

中国啮齿动物现生种类约210种，约占国内全部哺乳动物的41%[3]。因此，与一般中大型哺乳动物不同，出现在遗址中的啮齿目材料的种属更加复杂。要开展鉴定工作，首先要有较充分的分类学知识储备，理解啮齿目的基本特征、分类地位；重点掌握一批典型种属的头骨和牙齿的结构，弄清啮齿目鉴定应该重点观察的解剖部位。有了上述知识储备，才能够较顺利地阅读已有的啮齿目分类学专题文献，从中获取有用的描述与测量数据，再结合出土标本逐渐开展鉴定工作。下文将针对鉴定点做细致的讨论。

（一）啮齿目的基本特征

对于骨骼考古研究而言，啮齿目与哺乳纲真兽亚纲其他目相比较，其基本特征在于一套发达的啮合啃咬机构。首先，啮齿目具有发达的凿状门齿，无齿根，可终生生长。啮齿目具有啃咬物体的习性，以磨耗不断生长的门齿，维持正常的咬合状态。啮齿目犬齿缺失，在门齿与臼齿（或前臼齿）之间留有很宽的孔隙，称为"齿缺"，起到夹持容纳的功用（图版31：a）。其次，啮齿目的下颌骨坚固发达，但以弹性方式进行联结：下颌关节突呈狭长的弧背形（图版31：b），颞骨上对应的关节窝呈浅U形，两者构成松弛的结合；左右下颌骨的联合面也是弹性联结。这两处弹性联结使得啮齿目的上下颌形成了独特的啮合啃咬机构，下门齿可在上门齿后方与前方（图版31：c、d）进行啮合，也可以左右开合（图版31：e、f），实现啮取、咀嚼、研磨等功能，这是对粗糙植食纤维的进化适应。啮齿动物的上下颌骨在考古遗址常有发现，是重要的鉴定依据。

（二）分类地位

依据威尔逊《世界哺乳动物物种》第三版（2005）[4]，啮齿目共4个亚目：松鼠型亚目、鼠型亚目、豪猪型亚目和河狸型亚目。中国的啮齿动物在威尔逊新分类系统中划分为4个亚目、8个科，约82个属[5]。其中松鼠型亚目、鼠型亚目、豪猪型亚目是中国啮齿目最重要的三大亚目，而河狸型亚目在国内仅包含河狸一个种，本文不予专门讨论。

三大亚目的划分依据是深层和中层咬肌的结构位置关系和对应的骨骼架构关系[6]：松鼠型的眶前孔最小，深层和中层咬肌各自的一端都附着在颧弓上，另一端分别抵到下颌下缘和角突，眶前孔与咬肌附着无关（图1，a）。鼠型的眶前孔扩大，深层咬肌的一端附着在下颌下缘，另一端穿过眶前孔，附着在眶前孔的前方；中层咬肌一端附着在颧弓下方，一端在下颌角突上；眶前孔是咬肌的通道（图1，b）。豪猪型亚目的眶前孔最大，深层咬肌和中层咬肌各自的一端都附着在眶前孔中，另一端分别抵到下颌下缘和角突；眶前孔是咬肌附着位置（图2，c）。在鉴定实践中，可先依据上述特征对材料进行初步的分类，再做深入的检索。

图1　啮齿目三大亚目的中层及深层咬肌关系（深灰色块面示意眶前孔）[7]

Figure 1. Position of lateral and medial divisions of masseter muscle in three rodent suborders.

a. 松鼠型亚目　　b. 鼠型亚目　　c. 豪猪型亚目

a. Suborder Sciuromorpha.　　b. Suborder Myomorpha.　　c. Suborder Hystricomorpha.

五 头骨形态结构与鉴定要点

（一）头骨基本结构

啮齿目的头骨结构复杂，是分类鉴定的最重要依据。头骨由颅骨、下颌骨和舌骨三大部分构成。在考古遗址中，颅骨和下颌骨的残片常见于考古遗址，最具分类价值，是采集的重点；舌骨极难发现，分类意义不大，此处不予讨论。

啮齿目头骨的骨片可分为三大部分：咽颅部、脑颅部与下颌部。咽颅与脑颅以眼眶后缘的切线为界，合并为颅骨；下颌部仅包括左右下颌骨。啮齿目的头骨骨片结构详见表1，以黄胸鼠为例对其头骨主要骨片位置进行示意，见图版32：图1。

表1 啮齿目的头骨骨片结构一览表
Table 1. Bones of the skull in rodents.

部位	编号	骨片名称	数量	部位	编号	骨片名称	数量
咽颅部	1	鼻骨	1对	脑颅部	9	颞骨	1对
	2	前颌骨	1对		10	顶间骨	1块或无
	3	上颌骨	1对		11	枕骨	4块
	4	泪骨	1对		12	颧骨	1对
	5	腭骨	1对		13	前蝶骨	1块
	6	翼状骨	1对		14	基蝶骨	1块
脑颅部	无	犁骨	1块		15	听泡	1对
	7	额骨	1对或1块		16	乳骨	1对
	8	顶骨	1对	下颌部	17	下颌骨	1对

注：枕骨的4块骨片包括基枕骨1块、上枕骨1块、侧枕骨1对，但在早期已趋愈合；有些种（如中华鼢鼠）的顶间骨与其邻近的骨片愈合，不复存在；有些种（如长尾黄鼠）的1对额骨已经愈合成1块；犁骨位于鼻腔中央，构成鼻中隔的一部分，在图版32：图1中观察不到。

（二）头骨总体形态的观察

对于头骨形态，不仅要了解其基本的骨片结构，还要结合生态习性、栖息环境，观察其总体形态以及功能结构的特殊点，可从以下几方面着手观察。

1. 脑盒形态

脑盒形态即整个脑颅部所呈现的几何体形状，可结合颅顶和颅侧两个角度观察。常见脑盒结构可分为两大类。

高凸类：脑盒结构饱满，顶部凸出，脑颅部的相对体积较大，听泡大而圆、突出于颅底面，常见于营树栖或地表生活的松鼠类。包括圆球形，颅骨浑圆膨胀如球，如树栖的普通松鼠；扁球形，如地表活动的岩松鼠。

低平类：脑盒整体较扁平，顶部相对平坦，脑颅部的相对体积较小甚至小于咽颅部，听泡相对较扁，常见于各种穴居的鼠类。包括长扁椭圆形，如黑家鼠；短扁椭圆形，如小家鼠；方扁盒形，如青毛鼠；六边形扁形，如社鼠；后高前低的楔形，如甘肃鼢鼠。

2. 颧弓的粗细强弱、形态走向

颧弓与咬肌附着有关，不同种属的粗细强弱程度不一。左右颧弓还有平行、外突、内凹、正"八"字形、倒"八"字形等多种形态区分。同属不同种，其颧弓形态也有细微差异，可用于鉴定参考。以鼢鼠类为例：中华鼢鼠颧弓的后部比前部宽，或者前后宽窄相当（图版32：图2，a）；而东北鼢鼠的颧弓则是前宽后窄（图版32：图2，b），甘肃鼢鼠类似。再以竹鼠类为例：中华竹鼠颧弓的后部甚为宽展，约为颅长的75.8%~76.1%，左右颧弓近似等边三角形（图版32：图2，c）；银星竹鼠的颧弓则相对收敛（图版32：图2，d）；大竹鼠颧弓的前后相对较短，总体更为粗壮（图版32：图2，e）。

3. 吻部的长短宽窄

啮齿目的吻部是容纳门齿、实现啮合功能的关键结构之一，长短宽窄与其生态习性有关，可作为鉴定的参考。可用鼻骨前端到颧弓前缘的数据比较吻部的长短，用左右前颌骨的宽度比较吻部的宽窄。

（三）头骨上的脊、突、孔和窝

啮齿目头骨上有多种形态差异程度不一的脊、突、孔和窝等解剖结构，这些结构往往非常微小，容易被初学者所忽略，但却是啮齿目鉴定的关键所在。出土材料中难有完整的头骨，但颅骨残段往往会保留一些脊、突、孔、窝等结构，透过这些细微结构，配合牙齿等材料，往往能够鉴定到不同层次，甚至精确到种。如图版33所示，以多种啮齿动物为例，汇总常见的脊、突、孔、窝等解剖结构的位置，一些需要做重点的观察，一些可以作为描述或测量的定位点，需特别留意。以下对这些结构分别进行示例介绍。

1. 脊

成年个体的骨脊发育最为明显。肉眼可在家鼠大小的啮齿目头骨上观察到清晰的骨脊，用手抚摸有刮手感，可借助带标尺的放大镜进行观察与测量。

（1）眶上脊和顶脊

部分啮齿目在眼眶的上缘、往头后方向形成一条骨脊。一般从眶间最窄处开始形成一个尖缘，往头后逐渐明显并延伸到顶骨，然后逐渐尖灭；有的甚至在整个顶骨部分都很发达，往后延伸直到与侧枕骨相联。这条纵行的骨脊，属于额骨部分的称为"眶上脊"（图版33蓝标①），属于顶骨部分的称为"顶脊"或"颞脊"（图版33蓝标②），也有文献不加区分，将整个眶上脊和顶脊合称为"眶上脊"。

眶上脊和顶脊是啮齿目最重要的骨脊观察点，应该着重观察比较其有无、强弱、起始与消失位、走线形状等特征。

如青毛鼠与长尾巨鼠都属于大型鼠类，头骨尺寸和一些形态特征都很类似，但两种鼠类的眶上脊却有强弱之分：青毛鼠的眶上脊最宽处仅约0.35 mm，从颅顶观察可见与额骨交汇的折痕，但从颅侧方向观察却无明显折痕（图版34：图1，a）；长尾巨鼠的眶上脊最宽处可达 0.55 mm 左右，从颅侧观察可见眶上脊如同瓦沿突出于额骨（图版 34：图 1，b）。

又如褐家鼠和黑家鼠的眶上脊表现出起始与消失位、走线方向的差异：褐家鼠和黑家鼠的眶上脊都较发达，起始于眶间上缘，然后变粗并往后延伸。但褐家鼠的顶脊在顶骨上形成直线走向，左右顶脊构成明显的平行关系，且并不在顶骨后半部消失，而是最终与侧枕骨上的人字脊的左右部分相联接，形成一个骨脊整体。在大鼠属各种之中，只有褐家鼠的顶脊左右平行，这是褐家鼠独特的鉴定标示（图版 34：图 1，c）。而黑家鼠的左右顶脊在顶骨上呈对称的圆弧状，并在顶骨后半部逐渐消失（图版 34：图 1，d）。

再如大林姬鼠和小林姬鼠的头骨差异很小，但小林姬鼠无眶上脊（图版 34：图 1，e），而大林姬鼠却有眶上脊（图版 34：图 1，f），成为划分两个种的重要标示。

（2）人字脊和矢状脊

顶骨（或顶间骨）、颞骨与上枕骨、侧枕骨之间的骨脊即"人字脊"（图版 33 蓝标③）。两块顶骨之间的纵向骨脊即"矢状脊"（图版 33 蓝标④）。人字脊的强弱可作为比较的参考，如松鼠科的旱獭属（图版 34：图 2，a）、鼠科的竹鼠亚科（图版 34：图 2，b）都具有发达的人字脊和矢状脊；松鼠科黄鼠属（图版 34：图 2，c）的人字脊也较发达，却没有矢状脊。类似家鼠大小的啮齿目通常不具备发达的人字脊，但人字脊的左右部分稍发达，如鼢鼠的人字脊在脑顶不明显，但在左右枕骨部分却很明显（图版 34：图 2，d）。

2. 突

（1）角突、关节突和喙突

这三个突起（图版 33 紫标④⑤⑥）是下颌支末端上的显著突起，其形态和相对空间位置与咬肌的附着位置和走向有紧密的联系，具有重要的鉴定意义。出土材料常包含较多的下颌骨，但有时臼齿脱落。在这种情况下，可以通过观察这三个突，先对下颌进行初步分类，然后结合游离的臼齿材料做进一步的判断。

啮齿目三大亚目表现出各自的角突、关节突和喙突的特点：鼠型和松鼠型的喙突看上去难以区分；松鼠型的关节突显得更方整，顶部有结节（图版 35：图 1，a），鼠型的关节突更加圆钝（图版 35：图 1，b）；松鼠型的角突前方更往下延伸，形成一个"副突"（图版 35：图 1，a），而鼠型的角突往下方突出并不明显（图版 35：图 1，b）。豪猪型下颌支三个突起的特征更加明显：其喙突退化，几乎消失；关节突显著发达；角突呈较大的弧形结构（图版 35：图 1，c）。鼠型的竹鼠类与豪猪的角突类似，但尺寸明显要小，容易区分。

（2）门齿齿槽突

部分啮齿目的下颌骨颊侧后部有一个用于包纳下门齿末端的突起，如同刀鞘的末端，即所谓的"门齿齿槽突"（图版 33 紫标③）。松鼠型以及豪猪型啮齿目由于门齿末端不往颊侧生长，并不形成门齿齿槽突（如图版 35：图 1 中 a 为红松鼠，c 为中华豪猪，都无门齿齿槽突）。门齿齿槽突只见于鼠科、仓鼠科的部分种类，尤以竹鼠科的发达。另一些鼠科和仓鼠科种类的下门齿末端有突起感，但并不形成专门的鞘状齿槽突。

例如鼠科大鼠属的褐家鼠有明显的门齿齿槽突，但同为大鼠属的黑家鼠、黄胸鼠，以及小鼠属的小家鼠等，都没有形成门齿齿槽突。

（3）眶上突（眶后突）、眶前突

松鼠型啮齿目眼眶后上方的额骨延伸出一个特有的三角形突起，用以支护较大的眼球，通常称为"眶上突"，因为位置稍靠后，又称为"眶后突"（图版33 紫标⑧）。部分松鼠型啮齿目还有小的"眶前突"（图版33 紫标⑦）。

松鼠型啮齿目的两大族，即鼯鼠族和松鼠族，就是根据眶后突的形态差异加以区分的：鼯鼠族的眶上突向上翻翘，并带动眶上突的细长末端向上升起的角度更大（图版35：图2，a）；而松鼠族的眶上突仅水平方向往外延展，并不往上翻翘，眶上突的末端向上升起的角度也更小（图版35：图2，b）。

松鼠族各个种属之间眶上突的差异也较大。例如旱獭属眶上突厚实发达，表面还有细小的筛孔（图版34：图2，a）。再如岩松鼠的眶上突显得又薄又小（图版36：图1，a），红松鼠的眶上突则明显较大（图版36：图1，b）。

松鼠型啮齿目，如岩松鼠属、旱獭属和黄鼠属等，在眼眶前缘的额骨上有一个半圆形或圆形的小凹缺或小孔（即所谓的"上眶孔"），小孔后缘的小突起称为"眶前突"（图版33 紫标⑦）。花鼠的眶前突就较发达（图版36：图1，c）。另外一些松鼠型啮齿目，如丽松鼠属、花松鼠属、长吻松鼠属、巨松鼠属、多纹松鼠属都没有眶前突。

总之，眶上突和眶前突是松鼠型啮齿目最重要的鉴定标志点，需要特别注意。

（4）颧骨眶后突

部分啮齿目，主要是各种田鼠及䶄鼠，在颧骨前方偏内处，有一个小的突起，称为"颧骨眶后突"（图版33 紫标⑨），是识别田鼠类材料的重要标示，如中华䶄鼠、蒙古田鼠、社田鼠均有显著的颧骨眶后突（图版36：图2，b、c、d）。灰旱獭具有明显的"颧骨眶后突"（图版36：图2，a），是辨识灰旱獭的重要标示；但其他旱獭，如喜马拉雅旱獭等，则不具备该特征（图版34：图2，a）。

（5）副枕突与乳突

侧枕骨向下方伸出的尖状突起即是"副枕突"（图版33 紫标②）。侧枕骨紧邻乳骨，乳骨是听泡的附属骨骼。乳骨邻近副枕突的部分也往下延伸，称为"乳突"（图版33 紫标①）。乳突的突出程度明显不及副枕突。副枕突与乳突在鉴定时可起到参照作用。

（6）前颌骨额突

前颌骨末端嵌入颅背额骨的部分，通常称为"前颌骨额突"（图版33 紫标⑩）。不同啮齿目的前颌骨额突的伸入程度，以及与鼻骨末端构成的相对位置差异，都有一定的区别，可作为鉴定的参考特征。

3. 孔

（1）眶前孔

啮齿目在上颌骨颧弓突起的前方、吻部的两侧各有一个与眼眶相通的孔道，称为"眶前孔"（又称"眶下孔"，图版33 红标①）。眶前孔对啮齿目种属鉴定具有非常重要的意义。

松鼠型、鼠型和豪猪型啮齿目的眶前孔呈现出各自的特点：松鼠型的咬肌附着在颧弓上，眶前孔并不成为咬肌通道，因此穿孔很小，在孔的下方还有一个突起，用于附着咬肌（图版37：图1，a）；鼠型的深层咬肌穿过眶前孔，附着在眼眶的前方，因此穿孔较大（图版37：图1，b）；豪猪型的咬肌直接附着在眶前孔之中，因此穿孔显著扩大，几乎与眼眶等大（图版37：图1，c）。

鼠型啮齿目眶前孔的具体形态也呈多样化，有重要的鉴定意义。眶前孔的形态变化可以从颅顶、颅底和颅前三个方向进行观察比较。鼠型啮齿目眶前孔的典型形态有三种：

① 钥匙孔型——以鼠科最为典型。如褐家鼠的眶前孔（图版37：图2，a），从颅前观察，上大下窄，横短竖长，如同钥匙孔；从颅底观察，眶前孔位置有一个明显的缺口，这是供三叉神经通过的通道。各种鼠科动物该缺口形态有具体的差异，可用作鉴定的参考标准。

② 三角孔型——以鼢鼠科为典型。如中华鼢鼠的眶前孔（图版37：图2，b），从颅前观察，呈三角形，竖直长度稍大于横向长度；从颅底观察，没有供三叉神经通过的缺口，只有弧形的凹陷。

③ 横三角型——为竹鼠科动物所特有（图版37：图2，c），从颅前观察，呈三角形，长短径与前述两种相反，变成横长竖短；从颅底观察，颧板与吻侧完全愈合，没有供三叉神经通过的缺口。

（2）门齿孔

啮齿目在上门齿与上颊齿之间的硬腭上有一对狭长的孔，属于前颌骨的部分称为"门齿孔"，属于上颌骨的部分称为"颌孔"（也有文献称为"腭孔"）（图版33红标②）。除一些鼠兔属的啮齿目之外，其他啮齿目的这两种孔已经合为一孔，常统称为"门齿孔"。带有门齿孔的前颌骨残段在出土遗存中较常见，具有重要的鉴定意义。

对门齿孔的观察主要是比较其长短、宽狭，以及开孔末端与第一枚臼齿（或前臼齿）之间的距离。

松鼠型啮齿目，如岩松鼠的门齿孔（图版38：图1，d），均位于前颌骨，窄而短，前端靠近门齿，而末端远离第一枚前臼齿 P3 或 P4。

鼠科的门齿孔与上颌骨的颌孔已经合为一孔，通常显得窄而长。鼠科各种属的门齿孔距离臼齿 M1 有远有近。如褐家鼠与黑家鼠的门齿孔显得窄而长，几乎抵达臼齿 M1（图版38：图1，e、f）。

门齿孔距离臼齿 M1 的远近可以用于区分相近的种属。如大林姬鼠和小林姬鼠头骨形态非常相似，但小林姬鼠的门齿孔末端几乎与第一臼齿齐平（图版38：图1，a），而大林姬鼠的却有较远的距离（图版38：图1，b），可用这一特征将两个种区分开来。

鼢鼠科，如中华鼢鼠的门齿孔显得窄而短（图版38：图1，c），而仓鼠科的一些种类则显得更为宽大。

（3）腭孔、后鼻孔、枕大孔、颏孔、齿孔、上眶孔等

这些孔（图版33红标③~⑧）主要用作描述时的辅助定位，在此不作讨论。

4. 窝

（1）腭骨后侧窝

田鼠亚科中大多数的种在腭骨后缘突起两侧各发育有一个深窝，称为"腭骨后侧

窝"（图版33绿标③），是田鼠啮齿目的重要标示。

（2）硬腭窝

黄鼠属啮齿目的门齿孔前方、上门齿齿槽后方有一对硬腭窝（图版33绿标④），为该类动物所特有，具有鉴定意义。

（3）翼窝与翼内窝

主要用作描述时的辅助定位（图版33绿标①②），在此不作讨论。

六 门齿与颊齿

（一）门齿的基本特征

上门齿整体似半个圆环，下门齿则似四分之一的圆环。门齿齿冠如凿，前表面是较坚硬的釉质层（珐琅质层），其余部分为较易磨耗的齿质（白垩质）。门齿齿尖后缘有无缺刻、前表面有无纵沟、与上颌骨的夹角（垂直或前倾、后倾等）以及门齿的颜色（黄色、白色、橙色）等等，都是重要的分类鉴定观察点。如小家鼠的上门齿明显倾向后方，且上门齿齿尖后缘有一个台阶状的缺刻；而与小家鼠体型相当的小林姬鼠，门齿后缘则不具有缺刻（图版38：图2，a）。专营地栖生活的鼹形鼠为掏挖方便，其门齿向前显著伸突（图版38：图2，b）；长爪沙鼠的门齿前表面有一条纵沟（图版38：图2，c）。

（二）颊齿的基本特征

前臼齿和臼齿统称为"颊齿"。鼠型啮齿类一般缺乏前臼齿，松鼠型有1枚或2枚前臼齿，豪猪型有1枚前臼齿。各型啮齿类均有3枚臼齿。前臼齿和臼齿的区别，除了着生位置不同之外，还在于前臼齿咀嚼面上有突起，在幼年期有换齿现象；而臼齿的咀嚼面大、较平坦或有齿尖（齿突），但无换齿现象。

颊齿大体为长柱形，为了适应坚硬的谷物和其他植食，齿冠部的牙釉质伸入齿质部分，生成各种褶皱，使得咀嚼面形成各种异常复杂的纹路结构。啮齿目臼齿的咀嚼面样式繁多，其花纹结构是鉴定种属的重要标准，对比将另文专门讨论，这里仅简要介绍几种典型的啮齿目颊齿类型供一般分类参考。

岩松鼠属（图2，a）的上颊齿齿列由2枚前臼齿（P3与P4）、3枚臼齿（M1、M2、M3）所构成。P3退化变得很小，位于P4前方靠舌侧，咀嚼面为简单的单一齿突。P4已经臼齿化。P4、M1与M2均横向分成四行齿棱，中间两行显著突起，前后各一行较低矮；舌侧有一个显著的丘状突起。M3的咀嚼面为三角形，有两行齿棱，后侧为跟座。

大鼠属（图2，b）颊齿的咀嚼面呈三纵列的丘状齿突（结节），每一枚臼齿的咀嚼面又排成三行，在三行与三列的交点上，有的丘状齿突趋于发达，有的趋于退化甚至缺失。

板齿属（图2，c）的咀嚼面也分三纵列，但齿突在每一行上已经融合成板状，第一枚臼齿 M1 有三行，第二枚臼齿 M2 和第三枚臼齿 M3 各有两行。

仓鼠属（图 2，d）的咀嚼面为两个纵列的结节，左右对称，M1 有三行，M2 和 M3 有两行，且 M3 带跟部。

田鼠属（图 2，e）的咀嚼面也分为两个纵列，两个纵列均是牙釉质在臼齿的侧面楔入齿质形成的三角形齿棱，且两列齿棱左右交错排列，M3 带跟座。

鼢鼠属（图 2，f）的咀嚼面结构与田鼠属非常类似，也分两个纵列排列，靠近颊侧的齿棱形成三个（或两个）锐角形的齿页；M2 与 M3 靠近舌侧的齿棱各形成两个圆弧的齿页；M1 舌侧第一个齿页为锐角形，其余两个齿页为圆弧形。

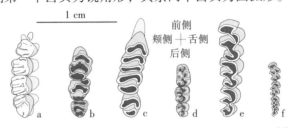

图 2　啮齿目的颊齿类型示例（均为左侧上颊齿齿列）[8]

Figure 2. Typical cheek types of rodents (left upper cheek teeth).

a. 岩松鼠属（岩松鼠）　b. 大鼠属（褐家鼠）　c. 板齿鼠属（印度板齿鼠）　d. 仓鼠属（大仓鼠）　e. 鼢鼠属（中华鼢鼠）　f. 田鼠属（普通田鼠）

a. *Sciurotamias davidianus*.　b. *Rattus norvegicus*.　c. *Bandicota indica*.　d. *Cricetulus triton*.　e. *Myospalax fontanieri*.　f. *Microtus avalis*.

七　头后骨骼

目前，除了少数几种啮齿动物，如豪猪、竹鼠、旱獭等，因为骨骼远远大于其他啮齿动物而能够被有效鉴定外，其他啮齿目由于体型相差不大，且前期研究工作不足，要进行比较精细的分类鉴定还有困难。

考古发掘出土的啮齿目头后骨骼以肩胛骨、肱骨、桡骨、尺骨、髋骨、股骨、胫骨和腓骨为主，另外有少量椎骨，其他部位的骨骼很难保存和被发现。鉴定时应当了解其出土细节，明确出土的头后骨骼与头骨遗存之间是否存有对应关系。在对应关系明确的情况下，可以通过头骨和牙齿遗存逐渐掌握一些重要种属（如几种家鼠）的骨骼特征，为更深入的鉴定工作做准备。头后骨骼鉴定的另外一个关键是将啮齿目的骨骼与体型相似的其他小哺乳动物（如食虫目、翼手目）区分开。限于篇幅，将另文讨论。

八　结语

啮齿目动物种属丰富、形态多样、尺寸微小，由于与之体态大小相似的还有食虫目、翼手目等多种动物，且考古遗址发掘出土的情况复杂多变，对啮齿目材料的收集和鉴定都是不小的挑战。但只要掌握了基本的观察点，鉴定工作就能够有的放矢，快速区分出土材料的大体类别。在此基础上再深入检索分类学文献，很多难题都能迎刃而解。希望本文能够给考古学同仁带来工作上的便利,同时能够加大相关材料的采集与鉴定力度,逐步揭示中国古人与啮齿目动物的独特关系。

致谢:文中岩松鼠、竹鼠、大林姬鼠等部分标本图片在河南省考古研究院得到侯彦峰先生支持并拍摄。中国社会科学院考古研究所袁靖先生对本文提出了修改意见。在此谨致谢意!

注释

[1] a. 袁靖:《中国动物考古学》,科学出版社,2015 年,第 283~284 页。

b. 武仙竹、袁东山:《黄胸鼠骨骼在古代城市考古中的发现》,《第四纪研究》2015 年第 3 期,第 784~786 页。

c. 武仙竹、王运辅:《小家鼠(Mus musculus)和褐家鼠(Rattus norvegicus)的化石材料与早期迁徙》,吉林大学边疆考古研究中心编《边疆考古研究(第 11 辑)》,科学出版社,2012 年,第 343~353 页。

[2] 尤玉柱:《史前考古埋藏学概论》,文物出版社,1989 年,第 5~89 页。

[3] 黄文几、陈延熹、温业新等:《中国啮齿类》,复旦大学出版社,1995 年,绪言页。

[4] Wilson D. E., Reeder D. A. M. (eds.), 2005. *Mammal Species of the World: A Taxonomic and Geographic Reference*. 3rd edition, Volume 1-2. Baltimore: The Jihns Hopkins University Press, pp. 185-211, 745-2142.

[5] 郑智民、姜志宽、陈安国编:《啮齿动物学》,上海交通大学出版社,2008 年,第 18~25 页。

[6] (美)E. H. 科尔伯特著,周明镇、刘后一、周本雄译:《脊椎动物的进化》,地质出版社,1976 年,第 329~333 页。

[7] 图引自(美)E. H. 科尔伯特著,周明镇、刘后一、周本雄译:《脊椎动物的进化》,图 101,地质出版社,1976 年,第 331 页。

[8] 图引自黄文几、陈延熹、温业新:《中国啮齿类》,图 3,复旦大学出版社,1995 年,第 4 页,有修改。

大数据及其在动物考古学中的应用*

中国新石器时代至青铜时代早期生业模式的特征

The Application of Metadata in Zooarchaeology
Evaluating Subsistence Patterns in Neolithic and Early Bronze Age China by Using Published Mammal Records

余翀

Chong Yu

中山大学社会学与人类学学院，广州

School of Sociology and Anthropology, Sun Yat-sen University, Guangzhou, China

摘要：对特定时空范围内的动物利用模式进行考察，其结果对理解生业模式转变、驯化动物的起源与传播等议题具有重要意义。对大量遗址的相关数据进行筛选、检验、变换以及运算，并借助特定的计算机技术、分析方法和展示手段，西方学者基于大数据在研究上述议题时提出了新的认识。本文简述了大数据的相关概念，并介绍了西方学者发表的两个实例，最后通过中国考古遗址出土哺乳动物骨骼的量化数据对新石器时代至青铜时代早期的生业模式进行探讨，希望能为我国学者开展综合研究提供借鉴与启发。

关键词：大数据；动物考古学；生业模式

Abstract: Investigating spatial and temporal variation in ancient mammal exploitation is essential for understanding the transition from hunting to agropastoralism and the dispersal of domestic animals. Addressing the role of metadata and its application in zooarchaeological studies in Southwest Asia and Europe, this paper explores dietary exploitation of mammals across the Chinese Neolithic and Early Bronze Age. The study period can be divided into four sections: the Early Neolithic ca. 10000 BP to 7500 BP; the Middle Neolithic ca. 7500 BP to 5000 BP; the Late Neolithic ca. 5000 BP to 4000 BP, and the Early Bronze Age ca. 4000 BP to 3600 BP. In total, I analyzed faunal assemblages from 103 sites/contexts, excluding sites whose total NISP is less than 10 or total MNI is less than 5.

---

* 本研究由国家社会科学基金青年项目（批准号18CKG005）资助。

In the Early Neolithic period, ruminants dominated in most of the sites, which may indicate that hunting was the main strategy and game was the source of dietary meat. However, some sites showed a considerable number of domestic pigs.

In the Middle Neolithic period, regional patterns predominated. In the middle Yellow River valley, the NISP of pigs increased dramatically-most of the sites in this region yielded a large proportion of pig bones, comprising up to 50 percent of the total NISP. Small mammals were the main dietary meat source in northeast China and north Shaanxi Province. The middle and lower Yangtze River valley and the south central region of Inner Mongolia contain relatively more bovids, particularly buffalos. Non-bovid ruminants, sheep and goat, were found across all sites and made up a considerable proportion of study assemblages. Medium-sized carnivores were also encountered very frequently, but in relatively low proportions. Pigs and ruminants were the main sources of dietary meat.

In the Late Neolithic period, pigs were still the dominant species among all mammal categories, comprising up to 50 percent of the total NISP and some of sites even reached 75 percent.

In the Early Bronze Age, sheep and goat mainly appeared in the west and central regions of the country. The time and the path of the spread of sheep and goat into East Asia are still in debate. More materials from Xinjiang Autonomous Region and the Hexi Corridor will be necessary for further study. The situation of cattle is more or less the same.

This paper provides a very preliminary metadata analysis of zooarchaeological assemblages, making use of GIS software. Further and refined results may be achievable by using advanced statistics, comparing data across a broader range, and combining databases from different regions.

Keywords: metadata, zooarchaeology, subsistence patterns

一 前言

中国的动物考古学研究始于20世纪30年代，至今已有公开发表的鉴定和研究报告200余篇及一些综合研究成果[1]。随着更多发掘工作的开展，以及动物考古学研究工作的深入，使得以单个或多个动物种属为研究对象的专题研究、特定时空范围内的动物种属频次变化研究和利用进阶统计分析方法进行的动物考古学综合研究成为可能与必然。关系模型数据库的应用，为学术资料的统一收集、整理和管理提供了科学的基础，同时极大地促进了学界的资源共享，也为上述研究的开展提供了先进、可靠的方法和手段[2]。但关系模型数据库的建立只是开展综合研究工作的第一步。因研究对象、目的和方法不同，需要将数据库内的数据进行筛选、检验、变换以及运算，这一过程需要借助特定的计算机技术、分析方法和展示手段——即大数据的应用。西方学者在大数据概念提出之初就将其应用到了动、植物考古学研究当中[3]。本文通过介绍西方学者基于大数据研究动物驯化的起源与传播的最新研究，希望为我们今后开展综合研究提供借鉴。

二 大数据概述

通常情况下，新生事物在其诞生的初始阶段并不会马上被严谨地定义，大数据也是如此。

由于计算机科学家察觉到传统的数据库定义及数据处理方式已经无法全面应对以前所未有的速度在不断增长和累积的数据，美国 META 集团（现 Gartner 集团）的分析师莱尼（Laney）在 2001 年将当时数据领域的复杂性总结为三个维度的变化，即数据的容量（Volume）、速度（Velocity）和种类（Variety），也被广泛地称为 3V；莱尼（Laney）还提出，大数据不仅仅是数据特性发生的变化，在数据处理与管理方面同样面临新的挑战[4]。3V 将大数据与仅具有 1V（Volume）的"海量数据"（Massive Data）与"超大规模数据"（Very Large Data）这些概念区分开来，成为用于描述大数据特征最为广泛的术语。

随后，矛若（Mauro）等人从数据、技术、方法和影响四个方面回顾并总结了大数据领域的理论背景，对大数据做出了一个较为全面的定义。文章认为，大容量、高速度和多种类是大数据信息的特征，信息技术与分析方法的革新是大数据得以发挥其价值的核心，而大数据的价值与影响则体现在将数据转换为对公司和社会有益的决策。基于此，大数据可以被定义为以大容量、高速度和多种类为特征的，需要运用特定的处理技术与分析方法将其价值进行转换的数据集合[5]。

三 大数据在西亚与欧洲动物考古学中的应用

大数据在动物考古学中最成功的应用当是由伦敦大学学院的申南（Shennan）教授主持的"近东地区及欧洲的家畜饲养的起源与传播"（The Origin and Spread of Stock-Keeping in the Near East and Europe，下文简称 OSSK）项目。该项目由英国艺术与人文研究委员会（Arts and Humanities Research Council）资助，旨在建立一个全面的考古遗址出土动物遗存数据库，并基于此开展关于近东地区及欧洲家畜饲养的起源与传播的大数据研究。该数据库包含了近东地区和欧洲所有已经发表的动物遗存（哺乳纲、鸟纲、鱼纲、软体动物以及甲壳纲）鉴定结果。种属鉴定结果依照发表时的内容进行输入，无论是鉴定到种、属、科，或者是"类别"（如山羊/绵羊），也有依身体尺寸进行分类的大、中、小型等。原文中的量化数据、测量数据、性别年龄比例、病理现象、骨骼部位出现频率、同位素以及基因检测的结果也都一并进行录入。

克罗尼（Conolly）等人在 OSSK 数据库中调取了 114 处位于亚洲西南部（约旦、叙利亚、以色列、巴勒斯坦、西奈半岛、伊朗、伊拉克、土耳其和塞浦路斯）和欧洲东南部（克里特岛、希腊和保加利亚）距今约 12000~7500 年的遗址出土的动物遗存数据，共计超过 40 万件骨骼标本[7]。量化分析的结果显示，动物资源的利用方式存在着明显的时空差异[6]。

幼发拉底河流域的遗址大都在距今10500年前后出现了家养动物,但家养动物所占的比例(可鉴定标本数,下同)都没有超过10%;距今9500年前后,家养动物的比例上升到40%左右;距今8500年前后,这个比例进一步上升到45%。在底格里斯河流域和扎格罗斯山区,早期家养动物的比例没有超过5%;至距今9500年前后,上升至20%;距今8400年前后,上升至40%。相比之下,立凡特地区的遗址出土的家养动物骨骼比例在较早的时期非常低,小于1%;到距今8800年前后,这一比例急速上升至平均35%左右。在立凡特地区的南部,家养动物的比例早期为1%,到距今8800年前后也仅有10%,表明家养动物传播到这一地区时并未较快地为当地居民所接纳。

在亚洲西南部的北部,家养动物的比例在距今10千纪的中晚期这一关键时期出现了大幅度的快速增长,并逐渐形成由绵羊、山羊、黄牛和家猪组成的家畜组合。这一现象为欧洲东南部距今9千纪的遗址的形成以及距今8千纪及其后的新石器聚落的出现提供了坚实的基础,由此往后,家养动物成为普遍存在。

曼宁(Manning)等人在OSSK数据库中调取了187处位于东南欧和中欧地区的遗址出土的动物遗存数据,这些遗址包含了自中石器时代晚期至新石器时代中期的240个考古学层位(距今约8000~4000年)。量化分析结果显示,动物资源的利用方式在这一区域同样存在着明显的时空差异[8]。

希腊和巴尔干地区新石器时代动物利用模式的差异较为显著,主要体现在山羊/绵羊和黄牛的比例上。在希腊地区,山羊/绵羊的比例(可鉴定标本数,下同)占77.5%,而黄牛只占5.9%。而在巴尔干地区,山羊/绵羊的比例占43.8%~53.8%,黄牛的比例则占到了25.7%~34.1%。这一现象的产生与两个地区的环境和气候有一定关系,相较而言,希腊的气候更为干燥,适合山羊和绵羊的生存;巴尔干地区温和的湿润气候适合喜欢茂密林木的黄牛和家猪。

在中欧,新石器时代属于Linearbandkeramik文化(简称LBK文化)的遗址出土的动物骨骼所反映的生业方式与希腊和巴尔干地区完全不同,即便家养动物和驯化作物从亚洲西南部起源再经由欧洲东部往西传播,最终到达中欧和西欧。这些生业模式与众不同的属于LBK文化的遗址集中在德国南部,在这些遗址中,考古学家发现了更多的黄牛、家猪以及野生动物(与东部地区相比)。有说法认为,出现这一现象是因为在农业人群到达中欧以前(进入新石器时代以前),这一地区仍属于中石器时代,生业模式依靠狩猎采集,生产力水平有限,人口密度相当低下,依靠种植谷物和豆类作物的农业人群向西移动至此,有了更多的机会和空间去开发野生资源。

四 中国新石器时代至青铜时代早期的生业模式

地理信息系统软件强大的计算和展示功能,可以科学直观地为我们提供不同动物的相对比例以及可能存在的在不同时空范围内的变化规律等信息。这些信息能够反映遗址(群)附近的自然环境、古代人类的生业模式、古代动物的消费模式,并为研究者提供关于古代人类对动物资源开发与利用方式特征的可比性数据。

据笔者统计，截至 2014 年，中国新石器时代至青铜时代早期考古遗址出土动物骨骼鉴定与研究报告中有关于哺乳动物数据的共计 199 处，其中有量化数据的 103 处，无量化数据的 96 处（表 1）。凡量化数据既有 NISP（可鉴定标本数）又有 MNI（最小个体数）的，选择使用 NISP 数据；若量化数据没有 NISP 只有 MNI 的，则选择使用 MNI 数据。NISP 总数小于 10 和/或 MNI 总数小于 5 的遗址归入无量化数据类别处理。

表 1　新石器时代至青铜时代早期考古遗址出土动物骨骼鉴定与研究报告中的遗址数量
Table 1. Numbers of mammal records from the Neolithic and Early Bronze Age sites, China.

年代	新石器时代早期	新石器时代中期	新石器时代晚期	青铜时代早期	合计
	距今 10000~7500 年	距今 7500~5000 年	距今 5000~4000 年	距今 4000~3600 年	
动物骨骼鉴定与研究报告中的遗址数量	24	81	68	26	199
有量化数据的遗址数量	10	46	30	17	103
无量化数据的遗址数量	14	35	38	9	96

建立关系模型数据库，将所有量化数据输入数据库，涵盖 91 个种属的哺乳纲动物，出土数量和肉量较小的（个体平均体重小于 1 公斤）啮齿目、食虫目和翼手目等的量化数据未用于统计分析。依据 73 个种属的哺乳纲动物分类学位置、出土数量、个体平均体重、生境以及可鉴定性的特征分为 7 大类：分别是其他大型哺乳纲、中型犬科、猪科、其他反刍亚目、牛族、小型哺乳纲、绵羊/山羊。

象科、大型食肉目、奇蹄目和骆驼科动物的平均体重均大于 100 千克，被归为"其他大型哺乳纲"。"中型犬科"主要指狗，也包括狼和豹，它们的个体体重相近，且后两者的出土数量不多。"猪科"包括野猪和家猪。"其他反刍亚目"包括偶蹄目中绝大部分种属，如麝科、鹿科以及牛亚科中除牛族、山羊和绵羊以外的所有种属。"小型哺乳纲"包括灵长目、大型啮齿目、大型兔形目、小型食肉目等个体体重不超过 30 千克的种属。"牛族"包括黄牛属和水牛属的现生和绝灭种属。"绵羊/山羊"则包含家养绵羊和山羊（附表 1）。

按以上分类方法对所有符合要求的数据进行检索和计算，得到新石器时代至青铜时代早期考古遗址出土各类哺乳动物的数量（和比例），再利用地理信息系统对数据进行组织和展示，得到各个时期考古遗址出土哺乳动物骨骼相对频率图（图版 39、40）。

在新石器时代早期，偶蹄目（其他反刍亚目和猪科）动物占所有遗址出土动物骨骼量的绝大多数。在这一时期，并没有动物考古学证据显示猪已经被驯化，获取动物性蛋白质的方式以狩猎为主。（图版 39：图 1，遗址编号、名称与参考文献见附表 2）

新石器时代中期，动物种类和比例出现了区域性分布差异。在黄河中游大多数遗址出土的哺乳动物骨骼中，猪科占有一半以上，表明猪科动物是古代居民的主要动物性蛋白质来源；东北和陕北地区古代居民多猎取并食用小型哺乳动物，这一现象或许与该地区的特殊生态环境有关；长江中游及下游地区、内蒙古中南部地区的大多数遗址较其他区域的遗址出土了更多的牛族骨骼；除牛族和绵羊/山羊以外的其他反刍亚目在这一时期的数量比例是最高的，而绵羊/山羊只在极少数遗址里少量出现；中型犬科动物的数量比例也较高，但其所占比重远不及猪科和其他反刍亚目动物。总体而言，猪科（主要是家猪）和其他反刍亚目仍然是肉食的主要供给者。（图版39：图2，遗址编号、名称与参考文献见附表3）

新石器时代晚期，猪科动物的比例仍是所有类别中最多的，其所占比例达到出土动物骨骼量一半以上的遗址数量相当多，甚至在不少遗址中能占75%以上。其余仍以其他反刍亚目动物为主。在这一时期，家养绵羊和山羊首次出现，主要分布在中西部地区。虽然我国关于家养绵羊和山羊的起源（来源）和传播问题现在尚未有定论，但是大数据研究表明这一时期的西北地区值得关注。（图版40：图1，遗址编号、名称与参考文献见附表4）

青铜时代早期，家养绵羊和山羊向东传播，相对比例明显增加，在黄河流域成为除家猪以外的重要家畜，狩猎获得的其他反刍亚目动物比例明显减小。长江流域及以南地区则仍保持以饲养家猪为主、狩猎为辅的动物性蛋白质获取方式。（图版40：图2，遗址编号、名称与参考文献见附表5）

五 结 语

已有的研究证实，山羊、绵羊、黄牛、家猪这四种家畜最早于距今1万年前后在立凡特地区和安那托利亚高原被驯化，在随后的数个千纪里，这些动物和栽培作物向西传播，距今约6000年到达了欧洲大陆的最西端和最北端[9]。但我们对动物驯化及其驯化完成后向外传播过程中的诸多细节仍不是很清楚。例如，西传的进程是持续性的还是有间断的？驯化物种和饲养技术的传播是同步的抑或是分离的？它们是经由人群的迁移一并移动的抑或是通过交换、贸易而传播的？家猪和黄牛有没有除西亚以外的其他独立起源中心（因为它们的野生祖本在欧亚大陆分布广泛），如有，它们与来自西亚的品种关系是怎样的？等等。

基于大数据的研究表明，在西亚及欧洲，家畜利用方式（主要体现在数量比例上）在不同的时间和空间（包括自然地理和考古学文化）范围内存在着明显的差异。可见家畜传播及开发利用的模式都具有多样性，其所反映的古代人群移动、交换贸易、食物加工、文化交流等方面的内涵比我们现有的认识要复杂得多。本文介绍的两例研究是动物驯化的起源与传播领域的最新成果，也是基于大数据在动物考古学应用中的最佳案例，希望能为我们在今后开展相关的综合研究提供启发与借鉴。

目前相关议题在中国动物考古学界的讨论相当有限，这与我们对基础数据的收集和管理一直沿用非标准化的手段，以及在对数据进行处理和计算时仅使用简单的算法有很大的关系。本文尝试使用已有的数据对新石器时代至青铜时代早期出土的各类动物的相对比例进行初步的考察，结果表明，自然环境差异、家养动物的出现和传播等与各类动物相对比例的区域性和历时性变化有着密不可分的关系。

注释

[1] 袁靖：《中国动物考古学文献目录》，《中国动物考古学》，文物出版社，2015年，第290~323页。

[2] 余翀：《关系模型数据库在动物考古学中的应用》，《考古》2015年第4期，第102~107页。

[3] a. Colledge S., Conolly J., Shennan S. 2004. Archaeobotanical evidence for the spread of farming in the Eastern Mediterranean. *Current Anthropology* 45(S4)：S35-S58.

b. Colledge S., Conolly J., Shennan S. 2005. The evolution of Neolithic farming from SW Asian origins to NW European limits. *European Journal of Archaeology* 8(2)：137-156.

c. Conolly J., Colledge S., Dobney K., et al. 2011. Meta-analysis of zooarchaeological data from SW Asia and SE Europe provides insight into the origins and spread of animal husbandry. *Journal of Archaeological Science* 38(3)：538-545.

d. Manning K., Stopp B., Colledge S., et al. 2013. Animal exploitation in the early Neolithic of the Balkans and central Europe, in Colledge S, Conolly J, Dobney K, et al. (eds.) *The Origins and Spread of Domestic Animals inSouthwest Asia and Europe*. Walnut Creek (CA)：Left Coast, pp. 237-252.

e. Manning K., Downey S., Colledge S., et al. 2013. The origins and spread of stock-keeping: the role of cultural and environmental influences on early Neolithic animal exploitation in Europe. *Antiquity* 87(338)：1046-1059.

[4] a. Laney D. 2001. 3-D data management: controlling data volume, velocity and variety. *META Group Research Note*.

b. https://blogs.gartner.com/doug-laney/files/2012/01/ad949-3D-Data-Management-Controlling-Data-Volume-Velocity-and-Variety.pdf.

[5] Mauro A. D., Greco M., Grimaldi M. 2016. A formal definition of big data based on its essential features. *Library Review* 65(3)：122-135.

[6] Conolly J., Colledge S., Dobney K., et al. 2011. Meta-analysis of zooarchaeological data from SW Asia and SE Europe provides insight into the origins and spread of animal husbandry. *Journal of Archaeological Science* 38(3)：538-545.

[7] Conolly J., Colledge S., Dobney K., et al. 2011. Meta-analysis of zooarchaeological data from SW Asia and SE Europe provides insight into the origins and spread of animal husbandry. *Journal of Archaeological Science* 38(3)：538-545.

[8] Manning K., Stopp B., Colledge S., et al. 2013. Animal exploitation in the early Neolithic of the Balkans and central Europe, in Colledge S, Conolly J, Dobney K, et al. (eds.) *The Origins and Spread of Domestic Animals inSouthwest Asia and Europe*. Walnut Creek (CA)：Left Coast, pp. 237-252.

[9] a. Peters J., Helmer D., von den Driesch A., et al. 1999. Early animal husbandry in the Northern

Levant. *Paléorient* 25(2): 27-48.

b. Hongo H., Meadow R. 1998. Pig exploitation at Neolithic Çayönü Tepesi (southeastern Anatolia). In: Nelson S. M. (ed.), *Ancestors for the Pigs: Pigs in Prehistory* (MASCA Research Papers in Science and Archaeology 15). Philadelphia: University of Pennsylvania Museum of Archaeology and Anthropology, pp. 77-98.

c. Peters J., von den Dreisch A., Helmer D. 2005. The upper Euphrates-Tigris basin: cradle of agro-pastoralism? In: Vigne J.-D., Peters J., Helmer D. (eds.), *The First Steps of Animal Domestication*. Oxford: Oxbow Books, pp. 96-124.

d. Zohary D., Hopf M., Weiss E. 2012. *Domestication of Plants in the Old World: The Origin and Spread of Domesticated Plants in Southwest Asia, Europe, and the Mediterranean Basin*. Oxford: Oxford University Press.

e. Helmer D., Gourichon L., Monchot H., et al. 2005. Identifying early domestic cattle from Pre-Pottery Neolithic sites on the Middle Euphrates using sexual dimorphism. In: Vigne J.-D., Peters J., Helmer D. (eds.), *The First Steps of Animal Domestication*. Oxford: Oxbow Books, pp: 86-95.

f. Hongo H., Pearson J., Öksüz B., et al. 2009. The process of ungulate domestication at Çayönü, Southeastern Turkey: a multidisciplinary approach focusing on *Bos* sp. and *Cervus elaphus*. *Anthropozoologica* 44(1): 63-78.

附表1 哺乳动物的分类

其他大型哺乳纲	亚洲象、虎、马、蒙古野驴、大熊猫、驴、棕熊、黑熊、貘、苏门犀、普氏野马、双峰驼
中型犬科	豹、狼、狗
猪科	野猪、家猪
其他反刍亚目	马麝、原麝、西伯利亚狍、马鹿、梅花鹿、毛冠鹿、麋鹿、黑麂、赤鹿、大角鹿、小鹿、白唇鹿、獐、鹅喉羚、黄羊、藏原羚、普氏原羚、北山羊、苏门羚、中华斑羚、盘羊、岩羊、水鹿
小型哺乳纲	短尾猴、猕猴、川金丝猴、藏酋猴、旱獭、河狸、豪猪、蒙古兔、草兔、野猫、猞猁、豹猫、花面狸、小灵猫、食蟹獴、豺、貉、沙狐、赤狐、水獭、猪獾、貂熊、紫貂、狗獾、鼬獾、黄鼬
牛族	黄牛、水牛、印度野牛、原始牛、圣水牛
绵羊/山羊	绵羊、山羊

附表2 新石器时代早期遗址编号、名称与参考文献

编号	遗址	参考文献
1	林西白音长汗	汤卓炜、郭治中、索秀芬：《白音长汗遗址出土的动物遗存》，内蒙古自治区文物考古研究所编《白音长汗》，科学出版社，2004年，第546~575页。
2	徐水南庄头	袁靖、李珺：《河北徐水南庄头遗址出土动物遗存研究报告》，《考古学报》2010年第3期，第385~391页。
3	济南月庄	宋艳波：《济南长清月庄2003年出土动物遗存分析》，北京大学考古文博学院编《考古学研究（7）》，科学出版社，2008年，第519~531页。

续附表 2

编号	遗址	参考文献
4	澧县八十垱	袁家荣：《动物遗骸》，湖南省文物考古研究所编《彭头山与八十垱》，科学出版社，2006 年，第 512～517 页。
5	澧县城头山	袁家荣：《城头山遗址出土的动物残骸鉴定》，湖南省文物考古研究所、国际日本文化研究中心编《澧县城头山》，文物出版社，2007 年，第 121～122 页。
6	邕宁顶蛳山	吕鹏：《广西邕江流域贝丘遗址的动物考古学研究》，中国社会科学院研究生院博士学位论文，2010 年。
7	商县紫荆	王宜涛：《紫荆遗址动物群及其古环境意义》，周昆叔主编《环境考古研究（第 1 辑）》，科学出版社，1991 年，第 96～99 页。
8	临潼白家村	周本雄：《白家村遗址动物遗骸鉴定报告》，中国社会科学院考古研究所编《临潼白家村》，巴蜀书社，1994 年，第 123～126 页。
9	宝鸡关桃园	胡松梅：《遗址出土动物遗存》，陕西省考古研究院、宝鸡市考古工作队编著《宝鸡关桃园》，文物出版社，2007 年，第 283～318 页。
10	秦安大地湾	祁国琴、林钟雨、安家瑗：《大地湾遗址动物遗存鉴定报告》，甘肃省文物考古研究所编《秦安大地湾》，文物出版社，2006 年，第 861～910 页。

附表 3　新石器时代中期遗址编号、名称与参考文献

编号	遗址	参考文献
1	白城双塔	张萌：《双塔遗址一期的动物利用方式研究》，吉林大学硕士学位论文，2011 年。
2	农安左家山	陈全家：《农安左家山遗址动物骨骼鉴定及痕迹研究》，吉林大学考古学系编《青果集》，知识出版社，1993 年，第 57～71 页。
3	敖汉赵宝沟	黄蕴平：《动物骨骼概述》，中国社会科学院考古研究所编《敖汉赵宝沟》，中国大百科全书出版社，1997 年，第 180～201 页。
4	凉城石虎山Ⅰ	黄蕴平：《石虎山Ⅰ遗址动物骨骼鉴定与研究》，内蒙古文物考古研究所、日本京都中国考古学研究会编《岱海考古（2）》，科学出版社，2001 年，第 489～513 页。
5	凉城王墓山坡上	内蒙古文物考古研究所、日本京都中国考古学研究会岱海地区考察队：《王墓山坡上遗址发掘报告》，内蒙古文物考古研究所、日本京都中国考古学研究会编《岱海考古（2）》，科学出版社，2001 年，第 200 页。
6	察右前旗庙子沟	黄蕴平：《庙子沟与大坝沟遗址动物遗骸鉴定报告》，内蒙古文物考古研究所编《庙子沟与大坝沟》，中国大百科全书出版社，2003 年，第 599～611 页。
7	察右前旗大坝沟	黄蕴平：《庙子沟与大坝沟遗址动物遗骸鉴定报告》，内蒙古文物考古研究所编《庙子沟与大坝沟》，中国大百科全书出版社，2003 年，第 599～611 页。

续附表3

编号	遗址	参考文献
8	林西白音长汗	汤卓炜、郭治中、索秀芬:《白音长汗遗址出土的动物遗存》,内蒙古自治区文物考古研究所编《白音长汗》,科学出版社,2004年,第546~575页。
9	林西井沟子西凉	陈全家:《林西县井沟子西凉新石器时代遗址出土动物遗存鉴定报告》,内蒙古自治区文物考古研究所、吉林大学边疆考古研究中心编《西拉木伦河流域先秦时期遗址调查与试掘》,科学出版社,2010年,第159~165页。
10	海拉尔哈克	黄蕴平、哈达:《动物遗存》,中国社会科学院考古研究所、内蒙古自治区文物考古研究所、内蒙古自治区呼伦贝尔民族博物馆、内蒙古自治区呼伦贝尔市海拉尔博物馆编《哈克遗址》,文物出版社,2010年,第190~200页。
11	垣曲古城东关	袁靖:《垣曲古城东关遗址出土动物骨骼研究报告》,中国历史博物馆考古部、山西省考古研究所、垣曲县博物馆编《垣曲古城东关》,科学出版社,2001年,第575~588页。
12	夏县辕村	安家瑗:《夏县裴介辕村遗址出土的动物骨骼》,《考古》2009年第11期,第24~25页。
13	常州圩墩	黄文几:《圩墩新石器时代遗址出土动物遗骨的鉴定》,《考古》1978年第4期,第241~243页。
14	沭阳万北	李民昌:《江苏沭阳万北新石器时代遗址动物骨骼鉴定报告》,《东南文化》1991年第3、4期,第183~189页。
15	高邮龙虬庄	李民昌:《自然遗物——动物》,龙虬庄遗址考古队编《龙虬庄》,科学出版社,1999年,第464~492页。
16	青浦崧泽	黄象洪、曹克清:《崧泽遗址中的人类和动物遗骸》,上海市文物保管委员会《崧泽》,文物出版社,1987年,第108~114页。
17	宁波傅家山	罗鹏:《傅家山遗址出土动物骨骼遗存鉴定和研究》,宁波市文物考古研究所、宁波市文物管理所编《宁波文物考古研究文集》,科学出版社,2008年,第61~73页。
18	丰都玉溪	赵静芳、袁东山:《玉溪遗址动物骨骼初步研究》,《江汉考古》2012年第3期,第103~112页。
19	濮阳西水坡	吕鹏、袁靖、杨梦菲:《西水坡遗址动物遗骸鉴定和研究》,南海森主编《濮阳西水坡》,中州古籍出版社、文物出版社,2012年,第659~696页。
20	郑州西山	陈全家:《郑州西山遗址出土动物遗存研究》,《考古学报》2006年第3期,第385~418页。
21	灵宝西坡	马萧林:《河南灵宝西坡遗址动物群及相关问题》,《中原文物》2007年第4期,第48~61页。
22	新安荒坡	侯彦峰、马萧林:《新安荒坡遗址出土动物遗存分析》,河南省文物管理局、河南省文物考古研究所编《新安荒坡》,大象出版社,2008年,第193~214页。
23	长阳桅杆坪	陈全家、王善才、张典维:《桅杆坪大溪文化遗址动物遗存研究》,陈全家、王善才、张典维著《清江流域古动物遗存研究》,科学出版社,2004年,第49~85页。

续附表3

编号	遗址	参考文献
24	长阳西寺坪	陈全家、王善才、张典维：《西寺坪大溪文化遗址动物遗存研究》，陈全家、王善才、张典维著《清江流域古动物遗存研究》，科学出版社，2004年，第85~102页。
25	长阳沙嘴	陈全家、王善才、张典维：《沙嘴大溪文化遗址动物遗存研究》，陈全家、王善才、张典维著《清江流域古动物遗存研究》，科学出版社，2004年，第102~116页。
26	宜都城背溪（南区）	李天元：《宜都城背溪遗址南区出土的动物遗存鉴定表》，湖北省文物考古研究所编《宜都城背溪》，文物出版社，2001年，第291页。
27	秭归柳林溪	武仙竹：《湖北秭归柳林溪遗址动物群研究报告》，国务院三峡工程建设委员会办公室、国家文物局编《秭归柳林溪》，科学出版社，2003年，第268~292页。
28	巴东楠木园	袁靖、杨梦菲、陶洋、罗运兵：《动物研究》，国务院三峡工程建设委员会办公室、国家文物局编《巴东楠木园》，科学出版社，2006年，第139~158页。
29	百色革新桥	宋艳波、谢光茂：《广西革新桥新石器遗址动物遗骸的鉴定与研究》，河南省文物考古研究所编《动物考古（第1辑）》，文物出版社，2010年，第218~231页。
30	邕宁顶蛳山	吕鹏：《广西邕江流域贝丘遗址的动物考古学研究》，中国社会科学院研究生院博士学位论文，2010年。
31	临潼姜寨	祁国琴：《姜寨新石器时代遗址动物群的分析》，西安半坡博物馆、陕西省考古研究所、临潼县博物馆《姜寨》，文物出版社，1988年，第504~538页。
32	商县紫荆	王宜涛：《紫荆遗址动物群及其古环境意义》，周昆叔主编《环境考古研究（第1辑）》，科学出版社，1991年，第96~99页。
33	临潼零口	张云翔、周春茂、阎毓民、尹申平：《陕西临潼零口村文化遗址脊椎动物遗存》，陕西省考古研究所编《临潼零口村》，三秦出版社，2004年，第525~533页。
34	靖边五庄果墚	胡松梅、孙周勇：《陕北靖边五庄果墚动物遗存及古环境分析》，《考古与文物》2005年第6期，第72~84页。
35	宝鸡关桃园	胡松梅：《遗址出土动物遗存》，陕西省考古研究院、宝鸡市考古工作队编著《宝鸡关桃园》，文物出版社，2007年，第283~318页。
36	高陵东营	胡松梅：《高陵东营遗址动物遗存分析》，陕西省考古研究院、西北大学文化遗产与考古学研究中心编《高陵东营》，科学出版社，2010年，第147~200页。
37	高陵杨官寨	胡松梅、王炜林、郭小宁、张伟、杨苗苗：《陕西高陵杨官寨环壕西门址动物遗存分析》，《考古与文物》2011年第6期，第97~107页。
38	华阴兴乐坊	胡松梅、杨岐黄、杨苗苗：《陕西华阴兴乐坊遗址动物遗存分析》，《考古与文物》2011年第6期，第117~125页。
39	商洛东龙山	胡松梅：《东龙山遗址动物遗存分析》，陕西省考古研究院、商洛市博物馆编《商洛东龙山》，科学出版社，2011年，第312~430页。
40	横山大古界	胡松梅、杨利平、康宁武、杨苗苗、李小强：《陕西横山县大古界遗址动物遗存分析》，《考古与文物》2012年第4期，第106~112页。
41	横山杨界沙	胡松梅、孙周勇、杨利平、康宁武、杨苗苗、李小强：《陕北横山杨界沙遗址动物遗存研究》，《人类学学报》2013年第32卷第1期，第77~92页。

续附表3

编号	遗址	参考文献
42	秦安大地湾	祁国琴、林钟雨、安家瑗：《大地湾遗址动物遗存鉴定报告》，甘肃省文物考古研究所编《秦安大地湾》，文物出版社，2006年，第861~910页。
43	礼县西山	余翀、吕鹏、赵丛苍：《甘肃省礼县西山遗址出土动物骨骼鉴定与研究》，《南方文物》2011年第3期，第73~79页。
44	同德宗日	安家瑗、陈洪海：《宗日文化遗址动物骨骼的研究》，河南省文物考古研究所编《动物考古（第1辑）》，文物出版社，2010年，第232~240页。
45	兴海羊曲十二档	安家瑗、陈洪海：《宗日文化遗址动物骨骼的研究》，河南省文物考古研究所编《动物考古（第1辑）》，文物出版社，2010年，第232~240页。
46	兴海香让沟	安家瑗、陈洪海：《宗日文化遗址动物骨骼的研究》，河南省文物考古研究所编《动物考古（第1辑）》，文物出版社，2010年，第232~240页。

附表4　新石器时代晚期遗址编号、名称与参考文献

编号	遗址	参考文献
1	伊金霍洛朱开沟	黄蕴平：《内蒙古朱开沟遗址兽骨的鉴定与研究》，《考古学报》1996年第4期，第515~536页。
2	房山镇江营和塔照	黄蕴平：《动物遗骸鉴定报告》，北京市文物研究所《镇江营与塔照》，中国大百科全书出版社，1999年，第557~565页。
3	侯马天马—曲村	黄蕴平：《天马—曲村遗址兽骨的鉴定和研究》，北京大学考古系商周组、山西省考古研究所编《天马—曲村》，科学出版社，2000年，第1153~1169页。
4	垣曲古城东关	袁靖：《垣曲古城东关遗址出土动物骨骼研究报告》，中国历史博物馆考古部、山西省考古研究所、垣曲县博物馆编《垣曲古城东关》，科学出版社，2001年，第575~588页。
5	襄汾陶寺	博凯龄：《中国新石器时代晚期动物利用的变化个案探究——山西省龙山时代晚期陶寺遗址的动物研究》，中国社会科学院考古研究所夏商周研究室编《三代考古（4）》，科学出版社，2011年，第129~182页。
6	兖州西吴寺	卢浩泉：《西吴寺遗址兽骨鉴定报告》，国家文物局考古领队培训班《兖州西吴寺》，文物出版社，1990年，第248~249页。
7	泗水尹家城	卢浩泉、周才武：《山东泗水县尹家城遗址出土动、植物标本鉴定报告》，山东大学历史系考古专业教研室编《泗水尹家城》，文物出版社，1990年，第350~352页。
8	兖州六里井	范春雪：《六里井遗址动物遗骸鉴定》，国家文物局考古领队培训班编著《兖州六里井》，科学出版社，1999年，第214~216页。
9	桓台前埠	宋艳波：《桓台唐山、前埠遗址出土的动物遗存》，山东大学东方考古研究中心编《东方考古（第5集）》，科学出版社，2009年，第315~345页。
10	苏州龙南	吴建民：《龙南新石器时代遗址出土动物遗骸的初步鉴定》，《东南文化》1991年第3、4期，第179~182页。

续附表 4

编号	遗址	参考文献
11	昆山绰墩	刘羽阳、袁靖：《绰墩遗址出土动物遗存研究报告》，苏州市考古研究所编《昆山绰墩遗址》，文物出版社，2011 年，第 372~380。
12	上海马桥	袁靖、宋建：《动物种属》，上海市文物管理委员会编著《马桥 1993~1997 年发掘报告》，上海书画出版社，2002 年，第 347~369 页。
13	蒙城尉迟寺	罗运兵、吕鹏、杨梦菲、袁靖：《动物骨骼鉴定报告》，中国社会科学院考古研究所、安徽省蒙城县文化局编《蒙城尉迟寺（第 2 部）》，科学出版社，2007 年，第 306~328 页。
14	茂汶营盘山	黄蕴平：《动物骨骼数量分析和家畜驯化发展初探》，河南省文物考古研究所编《动物考古（第 1 辑）》，文物出版社，2010 年，第 1~31 页。
15	濮阳西水坡	吕鹏、袁靖、杨梦菲：《西水坡遗址动物遗骸鉴定和研究》，南海森主编《濮阳西水坡》，中州古籍出版社、文物出版社，2012 年，第 659~696 页。
16	洛阳王湾	北京大学考古文博学院：《洛阳王湾》，北京大学出版社，2002 年，第 68、90 页。
17	登封王城岗	吕鹏、杨梦菲、袁靖：《动物遗骸的鉴定和研究》，北京大学考古文博学院、河南省文物考古研究所编著《登封王城岗考古发现与研究（2002~2005）》，大象出版社，2007 年，第 574~602 页。
18	禹州瓦店	吕鹏、杨梦菲、袁靖：《禹州瓦店遗址动物遗存的鉴定和研究》，北京大学考古文博学院、河南省文物考古研究所编者《登封王城岗考古发现与研究（2002~2005）》，大象出版社，2007 年，第 815~901 页。
19	新密新砦	黄蕴平：《动物遗存研究》，北京大学震旦古代文明研究中心、郑州市文物考古研究院编著《新密新砦》，文物出版社，2008 年，第 466~483 页。
20	渑池笃忠	杨苗苗、武志江、侯彦峰：《河南渑池县笃忠遗址出土动物遗存分析》，《中原文物》2009 年第 2 期，第 29~36 页。
21	黄梅塞墩	韩立刚：《黄梅县塞墩遗址动物考古学研究》，中国社会科学院考古研究所编《黄梅塞墩》，文物出版社，2010 年，第 329~346 页。
22	秭归庙坪	袁靖、孟华平：《庙坪遗址出土动物骨骼研究报告》，湖北省文物事业管理局、湖北省三峡工程移民局编《秭归庙坪》，科学出版社，2003 年，第 302~311 页。
23	巴东楠木园	袁靖、杨梦菲、陶洋、罗运兵：《动物研究》，国务院三峡工程建设委员会办公室、国家文物局编著《巴东楠木园》，科学出版社，2006 年，第 139~158 页。
24	临潼姜寨	祈国琴：《姜寨新石器时代遗址动物群的分析》，西安半坡博物馆、陕西省考古研究所、临潼县博物馆《姜寨》，文物出版社，1988 年，第 504~538 页。
25	商县紫荆	王宜涛：《紫荆遗址动物群及其古环境意义》，周昆叔主编《环境考古研究（第 1 辑）》，科学出版社，1991 年，第 96~99 页。
26	临潼康家	刘莉、阎毓民、秦小丽：《陕西临潼康家龙山文化遗址 1990 年发掘动物遗存》，《华夏考古》2001 年第 1 期，第 3~24 页。

续附表 4

编号	遗址	参考文献
27	高陵东营	胡松梅：《高陵东营遗址动物遗存分析》，陕西省考古研究院、西北大学文化遗产与考古学研究中心编著《高陵东营》，科学出版社，2010 年，第 147～200 页。
28	商洛东龙山	胡松梅：《东龙山遗址动物遗存分析》，陕西省考古研究院、商洛市博物馆编著《商洛东龙山》，科学出版社，2011 年，第 312～430 页。
29	永靖大何庄	中国科学院考古研究所甘青工作队：《甘肃永靖大何庄遗址发掘报告》，《考古学报》1974 年第 2 期，第 56 页。
30	西宁长宁	李谅：《青海省长宁遗址的动物资源利用研究》，吉林大学硕士学位论文，2012 年。

附表 5　青铜时代早期遗址编号、名称与参考文献

编号	遗址	参考文献
1	赤峰上机房营子	汤卓炜：《上机房营子遗址动物遗存初步分析》，内蒙古自治区文物考古研究所、吉林大学边疆考古研究中心编著《赤峰上机房营子与西梁》，北京，科学出版社，2012 年，第 249～252 页。
2	伊金霍洛朱开沟	黄蕴平：《内蒙古朱开沟遗址兽骨的鉴定与研究》，《考古学报》1996 年第 4 期，第 515～536 页。
3	夏县辕村	安家瑗：《夏县裴介辕村遗址出土的动物骨骼》，《考古》2009 年第 11 期，第 24～25 页。
4	泗水尹家城	卢浩泉、周才武：《山东泗水县尹家城遗址出土动、植物标本鉴定报告》，山东大学历史系考古专业教研室编《泗水尹家城》，文物出版社，1990 年，第 350～352 页。
5	昆山绰墩	刘羽阳、袁靖：《绰墩遗址出土动物遗存研究报告》，苏州市考古研究所编《昆山绰墩遗址》，文物出版社，2011 年，第 372～380 页。
6	上海马桥	袁靖、宋建：《动物种属》，上海市文物管理委员会编著《马桥 1993～1997 年发掘报告》，上海书画出版社，2002 年，第 347～369 页。
7	耿马石佛洞	何锟宇：《石佛洞遗址动物骨骼鉴定报告》，云南省文物考古研究所、中国社会科学院考古研究所、成都文物考古研究所、临沧市文物管理所、耿马傣族佤族自治县文化体育局编著《耿马石佛洞》，文物出版社，2010 年，第 354～363 页。
8	洛阳皂角树	袁靖：《古动物环境信息》，洛阳市文物工作队编《洛阳皂角树》，科学出版社，2002 年，第 113～119 页。
9	登封王城岗	吕鹏、杨梦菲、袁靖：《动物遗骸的鉴定和研究》，北京大学考古文博学院、河南省文物考古研究所编著《登封王城岗考古发现与研究（2002～2005）》，大象出版社，2007 年，第 574～602 页。
10	新密新砦	黄蕴平：《动物遗存研究》，北京大学震旦古代文明研究中心、郑州市文物考古研究院编《新密新砦》，文物出版社，2008 年，第 466～483 页。
11	偃师二里头	杨杰：《二里头遗址出土动物遗存研究》，中国社会科学院考古研究所编《中国早期青铜文化》，科学出版社，2008 年，第 470～539 页。
12	登封南洼	余翀：《动物遗骸》，郑州大学历史文化遗产保护研究中心编著《登封南洼》，科学出版社，2012 年，第 586～610 页。

续附表5

编号	遗址	参考文献
13	安阳鄣邓	侯彦峰、李素婷、马萧林、孙蕾：《安阳鄣邓遗址先商文化动物资源的获取与利用》，河南省文物考古研究所编著《安阳鄣邓》，大象出版社，2012年，第438~451页。
14	秭归官庄坪	武仙竹、周国平：《湖北官庄坪遗址动物遗骸研究报告》，国务院三峡工程建设委员会办公室、国家文物局编《秭归官庄坪》，科学出版社，2005年，第603~618页。
15	宜昌卜庄河	武仙竹：《三峡古动物序列》，《长江三峡动物考古学研究》，重庆出版社，2007年，第16~260页。
16	榆林火石梁	胡松梅、张鹏程、袁明：《榆林火石梁遗址动物遗存研究》，《人类学学报》2008年第27卷第3期，第232~248页。
17	商洛东龙山	胡松梅：《东龙山遗址动物遗存分析》，陕西省考古研究院、商洛市博物馆编著《商洛东龙山》，科学出版社，2011年，第312~430页。

后 记

　　本文集是 2016 年 10 月 13 日至 16 日郑州 "2016 国际动物考古协会理事会暨全球发展与中国视角动物考古学术研讨会" 的会后文集，是继 2014 年出版的《动物考古·第 2 辑》之后的第三本动物考古专辑。来自美国、加拿大、墨西哥、阿根廷、法国、德国、瑞典、土耳其、南非、澳大利亚、日本、印度和中国共 13 个国家的 18 位国际动物考古协会理事出席了理事会。来自美国哈佛大学、法国国家自然历史博物馆、北京大学、吉林大学、山东大学、中国科学院古脊椎与古人类研究所、中国社会科学院考古研究所、河南省文物考古研究院、陕西省考古研究院、湖北省文物考古研究所等 34 个海内外高校和科研院所的 44 位动物考古学者参加了学术研讨会。时任河南省文物局局长陈爱兰、副局长马萧林，时任河南省文物考古研究院院长贾连敏，以及国际动物考古协会副主席 Sarah Kansa 出席了会议并致辞。在此，对各位领导和专家的大力支持表示衷心的感谢！

　　本论文集的主编为河南博物院院长马萧林，客座主编为匈牙利中欧大学教授 Alice M. Choyke 和瑞典斯德哥尔摩大学教授 László Bartosiewicz。在本论文集的出版过程中，河南省文物考古研究院院长刘海旺给予了大力支持和指导，河南省文物考古研究院的王娟承担了与作者和出版社责任编辑的联络工作，在此一并表示诚挚感谢！

河南省文物考古研究院
2019 年 12 月

Figure 1. Late Chalcolithic stone structure at Godedzor (photograph by Alice Choyke).

Figure 2. Natural upright stone at center of Late Chalcolithic-Early Bronze age settlement with eastern mountain in background (photograph by Alice Choyke).

图版 1（Plate I）

From the Mountains to the Plains and Back Again

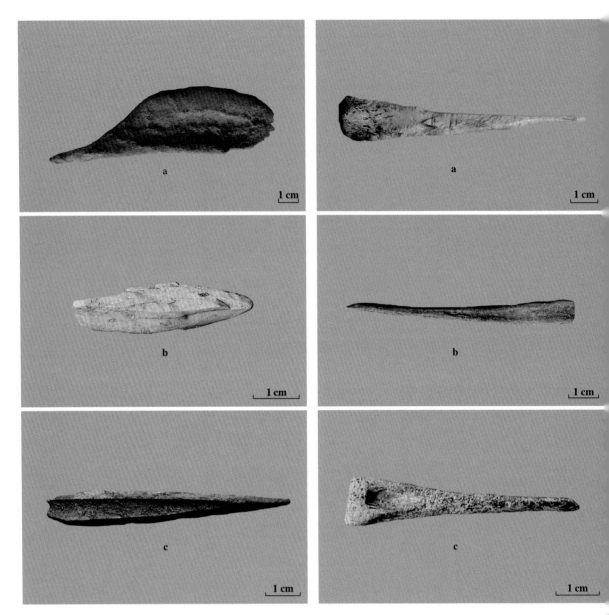

Figure 1. Barely modified Class II and I–II pointed tools made from unselected raw materials (Photograph by Alice Choyke).
a. Class II awl. b. Class II awl 2. c. Class I to II awl.

Figure 2. Planned Class I pointed tools made from selected raw materials (Photograph by Alice Choyke).
a. planned groove and split metapodial b. planned groove and split metapodial proximal. c. planned groove and split metatarsal awl.

From the Mountains to the Plains and Back Again

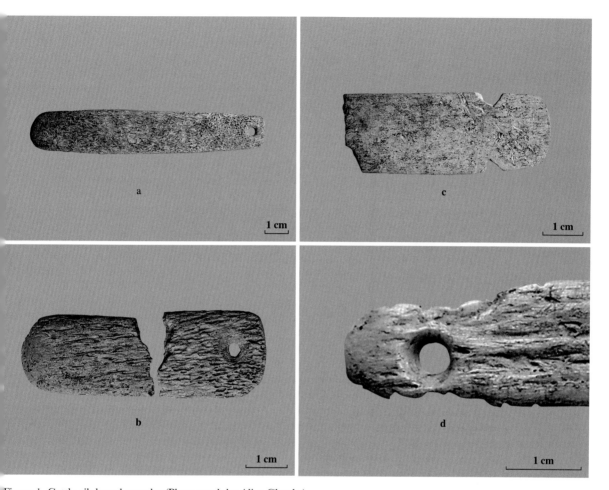

Figure 1. Cattle-rib based spatulas (Photograph by Alice Choyke).
a. rib based spatula with perforation 2.   b. rib based spatula with perforation.
c. rib based notched spatula without perforation.   d. notched perforated head closeup.

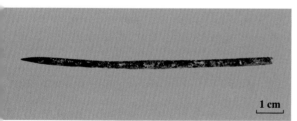

Figure 2. Planned fine bone needle made from longitudinal long bone splinter (Photograph by Alice Choyke).

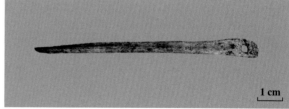

Figure 3. Nahl-binding (single-needle knitting) needle made from domestic caprine sized tibia diaphysis fragment (Photograph by Alice Choyke).

From the Mountains to the Plains and Back Again

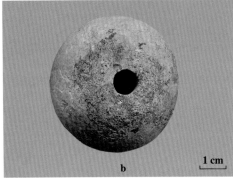

Figure 1. Spindle whorls used in spinning made from the sawn off femur caput of cattle, red deer and aruochs (Photograph by Alice Choyke).
a. spindle whorl.
b. spindle whorl 2.

Figure 2. Red deer antler weaving comb with incised decoration with evidence of long-time use (Photograph by Alice Choyke).

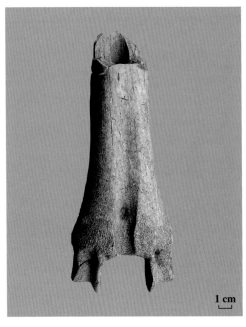

Figure 3. Cattle metatarsal with modified distal end of unknown function but with broad distribution geographic and temporal (Photograph by Alice Choyke).

Figure 4. Smoothed interior of notch on distal end of cattle metatarsal (Photograph by Alice Choyke).

Figure 5. Less well preserved cattle metatarsal with modified distal end of unknown function (Photograph by Alice Choyke).

图版 4（Plate IV）

From the Mountains to the Plains and Back Again

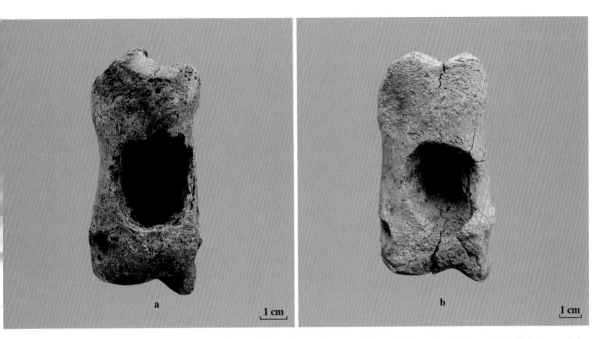

Figure 1. Unfinished perforated first phalanges from wild or domesticate cattle, possibly used as light weights (Photograph by Alice Choyke).
a. unfinished perforated cattle first phalange.   b. unfinished perforated cattle first phalange 2.

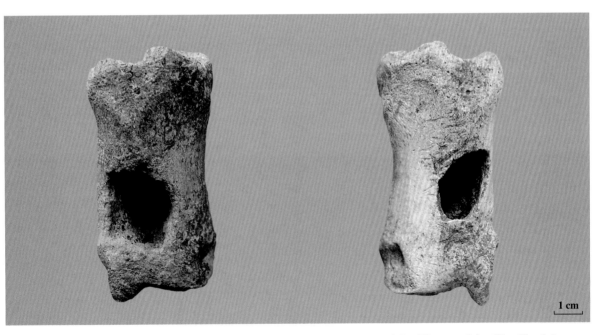

Figure 2. Two perforated first phalange of domestic cattle, possibly used as light weights (Photograph by Alice Choyke).

图版 5（Plate V）

From the Mountains to the Plains and Back Again

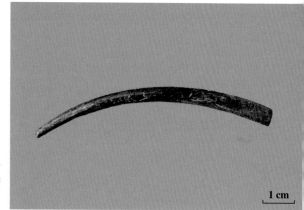

Figure 1. Thin, finely made, pointed tusk tool of either special utilitarian or ornamental function (Photograph by Alice Choyke).

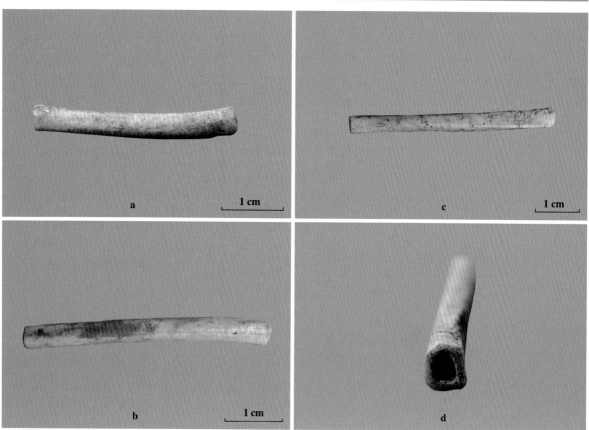

Figure 2. Long bone beads/or clasp elements made mostly from metapodial diaphyses of small canids and hare (Photograph by Alice Choyke).
a. small canid metapodial bead.   b. hare metapodial bead lengthwise.   c, d. hare metapodial bead 2.

From the Mountains to the Plains and Back Again

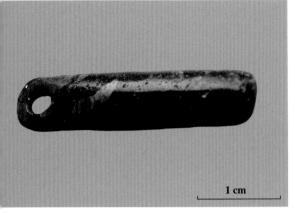

Figure 1. Cylindrical pendant/bead made from thick cortical long bone and burned black (Photograph by Alice Choyke).

Figure 2. Perforated brown bear canine, artifically flattened on one surface to lie flat (Photograph by Alice Choyke).

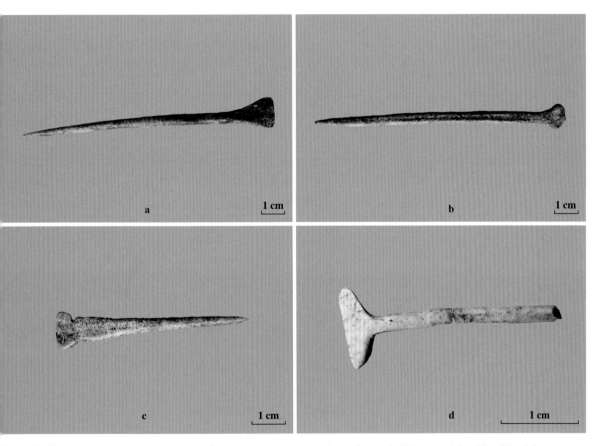

Figure 3. Home-made decorative pins made from large ungulate long bone diaphysis (Photograph by Alice Choyke).
a. homemade decorative pin 1.   b. homemade decorative pin 3.   c. homemade decorative pin curated.   d. homemade decorative pin 2.

图版 7（Plate VII）

From the Mountains to the Plains and Back Again

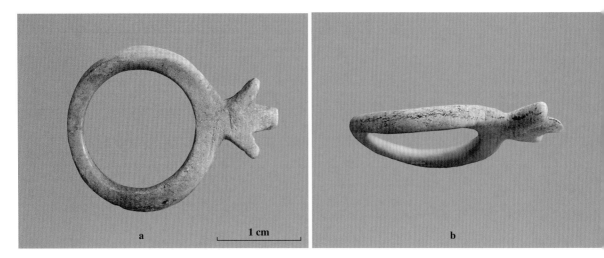

Figure 1. Ring or clasp element made possibly from a wild caprine long bone diaphysis (Photograph by Alice Choyke).
a. ring front.   b. curved ring.

Figure 2. Pigment container probably made from bezoar goat femur diaphysis with densely incised ornamentation (Photograph by Alice Choyke).

Middle Bronze to Early Iron Age Bone Tools from Eastern Germany

Figure 1. Bone tool no.1 from Kemnitz (BLADAM 2001–779/18/3/11; Photos: Teegen).
a. complete.   b. abrasion marks.

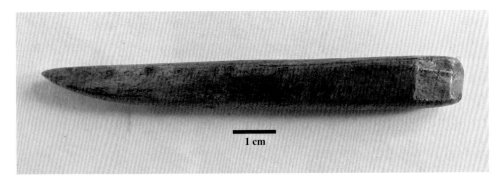

Figure 2. Bone tool no. 2 from Kemnitz (BLADAM 2001–779/38/3/24; Photo: Teegen).

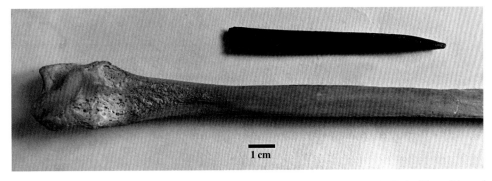

Figure 3. Bone tool no. 3 from Kemnitz (BLADAM 2001–779/53/3/46) and left pig fibula (Photo: Teegen).

Middle Bronze to Early Iron Age Bone Tools from Eastern Germany

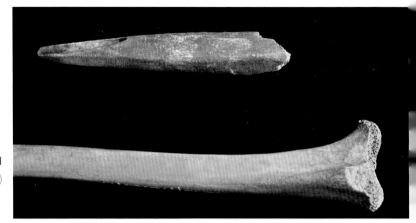

Figure 1. Spearhead from Kemnitz (tool no. 4, BLADAM 2001-779/41/3/34) and right sheep tibia (Photo: Teegen).

Figure 2. Spearheads from Hjortspring, Denmark (from Kaul; Photo: Larsen).

Middle Bronze to Early Iron Age Bone Tools from Eastern Germany

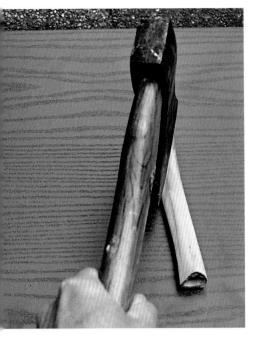

Figure 1. Replica of a projectile made from a left sheep tibia by a left-handed person (Photo: Küchelmann).

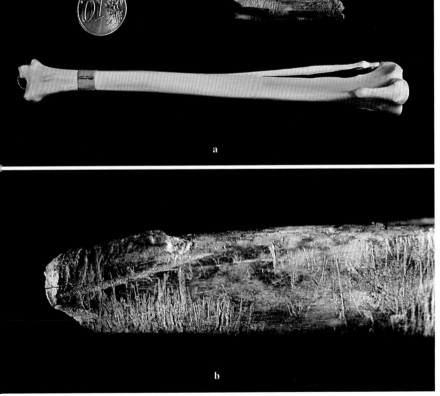

Figure 2. Worked hare tibia fragment from Kemnitz (artefact no. 5; BLADAM 2001–779/38/3/23; Photos: Teegen).
a. artefact with right tibia of a hare.
b. detail of abrasion marks.

Maintenance, Inheritance and Memory

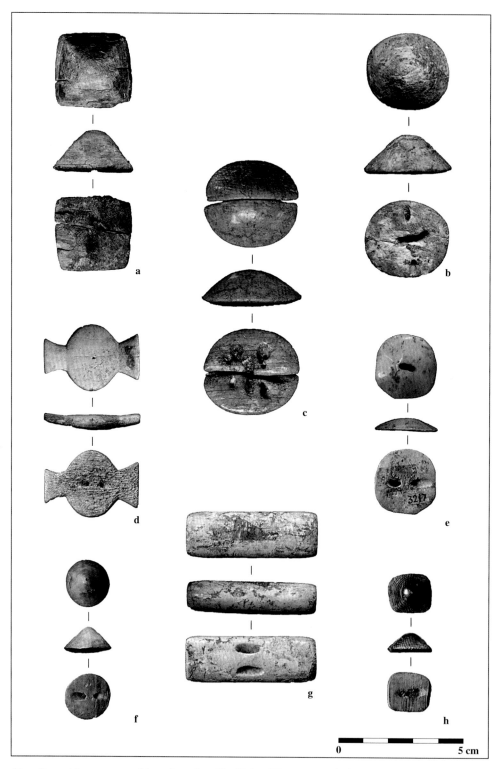

V-perforated buttons assemblage from Los Castillejos. They are made from elephant ivory (a–c, e–h) and red deer antler (d).

图版 12（Plate XII）

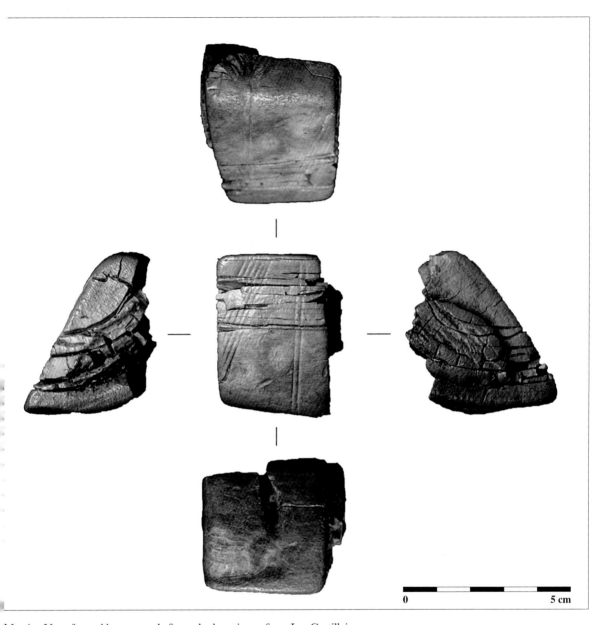

Massive V-perforated button made from elephant ivory, from Los Castillejos.

图版 13（Plate XIII）

Maintenance, Inheritance and Memory

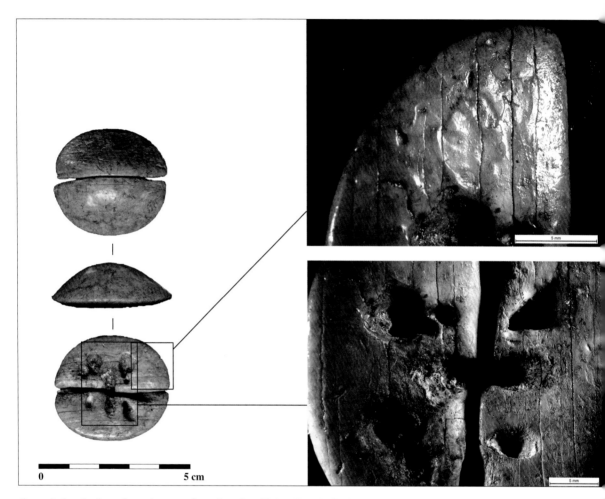

One of the elephant ivory buttons from Los Castillejos. Detailed pictures of the extremelly worn surface and repeated curations of the broken V-perforations.

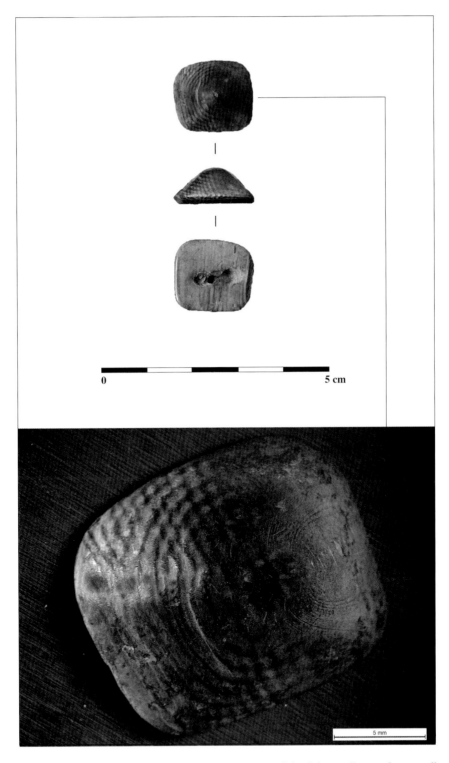

One of the elephant ivory buttons from Los Castillejos. Detailed picture of the Schreger lines and extremelly worn surface.

图版 15（PlateXV）

The Fauna from Lion Island, a Late Nineteenth and Early Twentieth Century Chinese Community in British Columbia, Canada

Figure 1. Remains of the Ewen Cannery and the western tip of Lion Island (Photo by D. Ross).

Figure 2. Excavation of artefacts from the Chinese bunkhouse on Lion Island (Photo by D. Ross).

Figure 3. Pig bone recovered from the Chinese bunkhouse on Lion Island. Trowel is 24 cm long (Photo by D. Ross).

吉林大安后套木嘎遗址出土贝类遗存研究
The Study of Shell Remains from the Houtaomuga Site in Da'an, Jilin

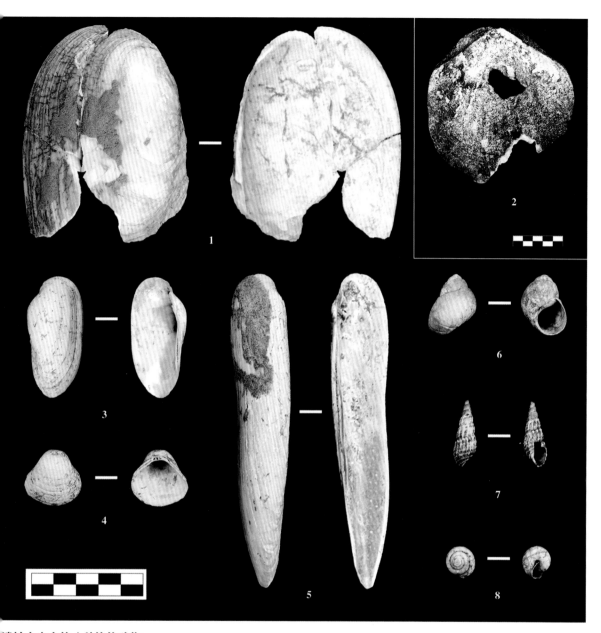

遗址中出土的八种软体动物

Eight taxa of mollusk unearthed from HTMG.

1. 无齿蚌亚科（左壳, 11DHAⅢT1006③：1） 2. 虾夷盘扇贝（左壳, 111DHAⅢT1305②：7） 3. 圆顶珠蚌（右壳, 114DHAIF25①：3）
4. 河蚬（右壳, 115DHAIH295：2） 5. 短褶矛蚌（左壳, 112DHAⅢG18：434） 6. 中国圆田螺（12DHAⅢG21：79） 7. 纵肋织纹螺（11DHAⅢH92：348） 8. 灰巴蜗牛（11DHAⅢG3：80）

图版 17（Plate XVII）

吉林大安后套木嘎遗址出土贝类遗存研究
The Study of Shell Remains from the Houtaomuga Site in Da'an, Jilin

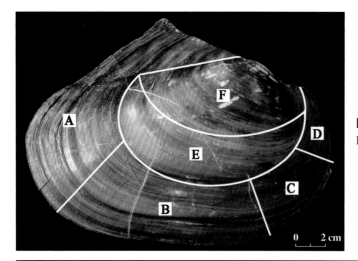

图 1　无齿蚌亚科壳体分区
Figure 1. Zones of the Anodontinae shell.

图 2　无齿蚌亚科蚌料典型标本
Figure 2. Typical specimens of semi-finished products of Anodontinae shells.
1. Ⅰ型蚌料（13DHAIG2：13） 2. Ⅱ型蚌料（11DHAⅢH25：2） 3. Ⅱ型蚌料（11DHAⅢH92：25） 4. Ⅳ型蚌料（11DHAⅢH92：164）
5. Ⅴ型蚌料（14DHAIH175：1） 6. Ⅵ型蚌料（12DHAIVT0609①：10） 7. Ⅶ型蚌料（11DHAⅢ①：21） 8. Ⅷ型蚌料（12DHAIH166：1）
9. Ⅸ型蚌料（11DHAⅢF1：2） 10. Ⅹ型蚌料（12DHAⅢH172：1） 11. Ⅺ型蚌料（12DHAⅢG22：27） 12. Ⅻ型蚌料（12DHAⅢH92：5）

吉林大安后套木嘎遗址出土贝类遗存研究
The Study of Shell Remains from the Houtaomuga Site in Da'an, Jilin

图1　锋利工具来回划动形成的痕迹

Figure 1. Traces formed by a sharp tool sliding back and forth.

图2　剔刮法加工形成的平齐茬口

Figure 2. Fracture surface formed by scraping and percussion.

图3　I型蚌料与穿孔蚌刀

Figure 3. Semi-finished product and finished perforate shell knives.

1. I型蚌料（13DHAIG2：13）　2. AⅢ型蚌刀 13DHAIF6①：1）　3. AⅡ型蚌刀（13DHAIF4①：12）　4. BI型蚌刀（13DHAⅢT1309②：3）　5. BⅡ型蚌刀（11DHAⅢF1：1）　6. AI型蚌刀（13DHAIF4①：11）

图版 19（Plate XIX）

吉林大安后套木嘎遗址出土贝类遗存研究
The Study of Shell Remains from the Houtaomuga Site in Da'an, Jilin

XI 型蚌料与蚌匙

Semi-finished products and finished shell spoons.

1. Ⅲ型蚌料（11DHAⅢH92：25） 2. A型蚌匙（11DHAⅢT1512②：2） 3. Ⅳ型蚌料（11DHAⅢH92：164）
4. B型蚌匙（11DHAⅢT1512②：1）

吉林大安后套木嘎遗址出土贝类遗存研究
The Study of Shell Remains from the Houtaomuga Site in Da'an, Jilin

蚌片加工工艺流程图
Operational chain of shell tablets.

1. IX型蚌料（11DHAIIIF1：2） 2. B型蚌片半成品（12DHAIIIF8：13） 3. B型蚌片（13DHAIF6①：38） 4. AI型蚌片半成品 12DHAIIIG18：2）
5. AI型蚌片半成品（11DHAIIIH92：1） 6. AI型蚌片半成品（11DHAIIIM30：2） 7. AI型蚌片（12DHAIIIG18②：2） 8. XI型蚌料
（12DHAIIIG22：27） 9. VI型蚌料（12DHAIVT0609①：10） 10. VIII型蚌料（12DHAAIIIH166：1） 11. X型蚌料 12DHAIIIH172：1）
12. V型蚌料（14DHAIH175：1） 13. VII型蚌料（11DHAIII①：21） 14. AII型蚌片半成品（11DHAIIIH92：114） 15. AII1型蚌片
（11DHAIIIT1412②：4） 16. AII2型蚌片（11DHAIIIT1305②：3） 17. AII3型蚌片（11DHAIH92：4）

The Study of Shell Remains from the Houtaomuga Site in Da'an, Jilin

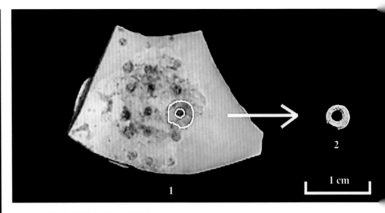

图 2 蚌珠毛料取料示意图
Figure 2. Operational chain of shell beads.
1. 蚌珠毛料（11DHAIIIT1005①：1） 2. 蚌珠

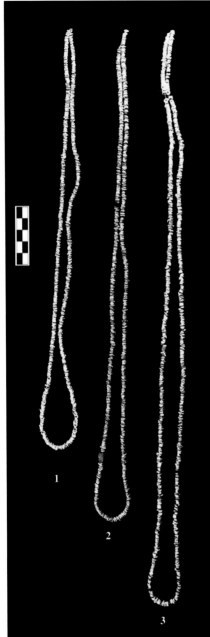

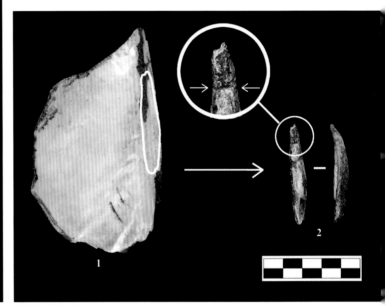

图 1 蚌珠串饰
Figure 1. Strings of shell beads.
1. 12DHAIIIM91：1
2. 12DHAIIIM92：1
3. 12DHAIIIM79：1

图 3 三棱形蚌饰取料示意图
Figure 3. Operational chain of the triangular shell ornament.
1. IV型蚌料（11DHAIIIH92：164） 2. 三棱形坠饰（11DHAIIIH85：1）

吉林大安后套木嘎遗址出土贝类遗存研究
The Study of Shell Remains from the Houtaomuga Site in Da'an, Jilin

图1 矛形蚌饰加工流程示意图

Figure 1. Operational chain of the spearhead-shaped shell ornament.

1. XII 型蚌料（12DHAIIIH92：5）
2. 长方形厚料块（11DHAIIIT1405②：2）
3. 矛形蚌坠饰（12DHAIIIM90：2）

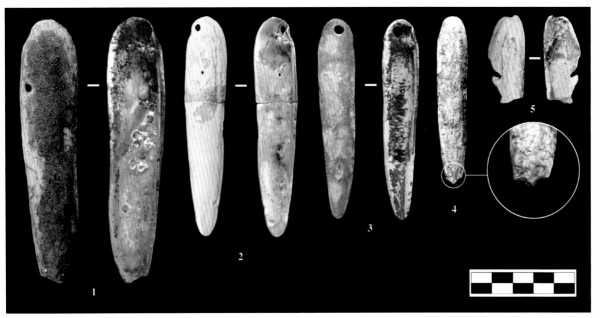

图2 短褶矛蚌制品

Figure 2. Shell artifacts made of *Lanceolaria glayana*.

1. A 型穿孔蚌壳（11DHAIIIT1205③：34） 2. BI 型穿孔蚌壳（11DHAIIIM58：1） 3. BII 型穿孔蚌壳（11DHAIIIM16：14）
4. 刻划器（11DHAIIIH89：B15） 5. 磨制蚌片（13DHAIH47：25）

吉林大安后套木嘎遗址出土贝类遗存研究
The Study of Shell Remains from the Houtaomuga Site in Da'an, Jilin

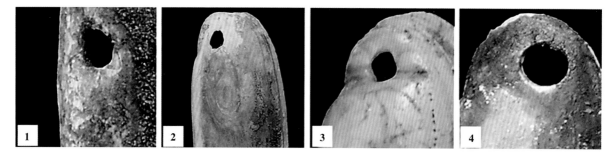

图1 四种穿孔方式留下的痕迹
Figure 1. Four styles of perforation on *Lanceolaria glayana* shells.
1. 直接打制（11DHAⅢT1205③∶34） 2. 预制台面后砸击（11DHAⅢM67∶1） 3. 锯割（11DHAⅢM46∶1）
4. 管钻（11DHAⅢM16∶14）

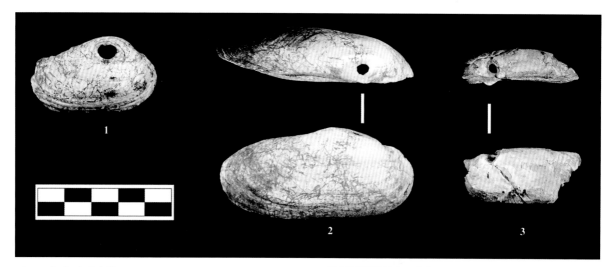

图2 穿孔圆顶珠蚌
Figure 2. Perforated *Unio douglasiae* shells.
1. 直接打制穿孔蚌壳（11DHAⅢH91∶10）
2. 预制台面后砸击穿孔蚌壳（11DHAⅢT1312③∶3）
3. 锯割穿孔蚌壳（11DHAⅢH92∶213）

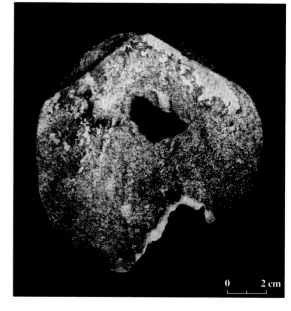

图3 虾夷盘扇贝制品（12DHAⅢT1305②∶7）
Figure 3. Shell artifact made of *Mizuhopecten yessoensis*.

图版 24（Plate XXIV）

黄牛与牦牛骨骼形态的对比观察
Differences in Osteological Morphology between Cattle (*Bos taurus*) and Yak (*Bos grunniens*)

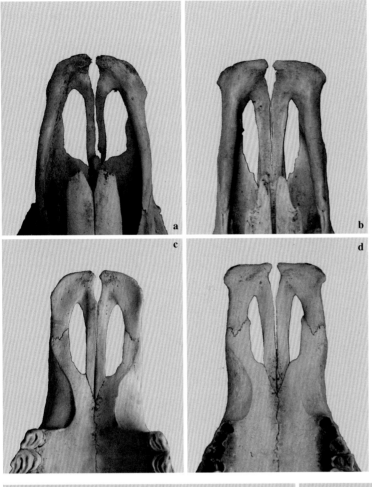

图 1　黄牛、牦牛的前颌骨
Figure 1. Premaxilla of cattle and yak.
a. 黄牛（背侧观）
　cattle (dorsal view).
b. 牦牛（背侧观）
　yak (dorsal view).
c. 黄牛（腹侧观）
　cattle (ventral view).
d. 牦牛（腹侧观）
　yak (ventral view).

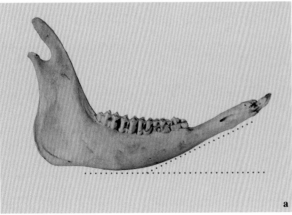 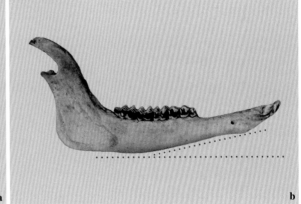

图 2　黄牛、牦牛的下颌骨（侧面观）
Figure 2. Mandible of cattle and yak (lateral view).
a. 黄牛　　b. 牦牛
　cattle.　　yak.

图版 25（Plate XXV）

黄牛与牦牛骨骼形态的对比观察
Differences in Osteological Morphology between Cattle (*Bos taurus*) and Yak (*Bos grunniens*)

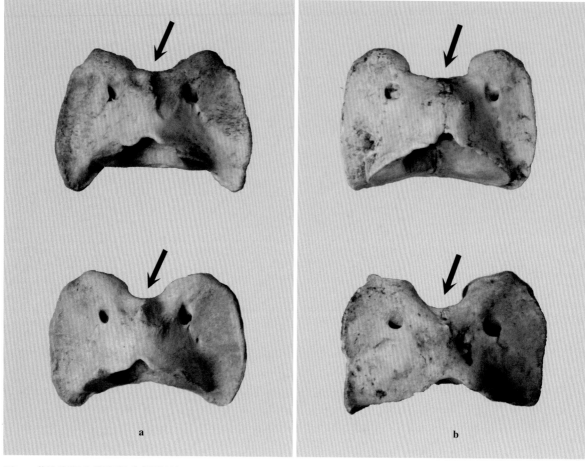

图 1　黄牛和牦牛的寰椎（背侧观）　　　　a. 黄牛　　b. 牦牛
Figure 1. Atlas of cattle and yak (dorsal view).　　　cattle.　　yak.

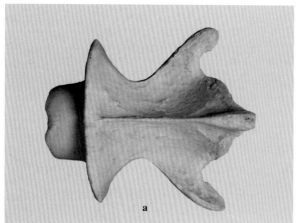 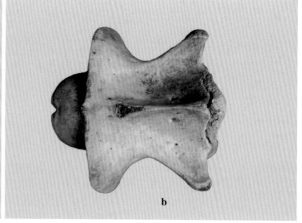

图 2　黄牛和牦牛的枢椎（腹侧观）　　　　a. 黄牛　　b. 牦牛
Figure 2. Axis of cattle and yak (ventral view).　　　cattle.　　yak.

黄牛与牦牛骨骼形态的对比观察
Differences in Osteological Morphology between Cattle (*Bos taurus*) and Yak (*Bos grunniens*)

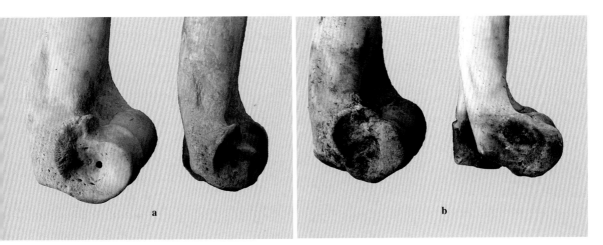

图1 黄牛和牦牛的肱骨远端（外侧观） a. 黄牛 b. 牦牛
Figure 1. Distal end of cattle and yak humeri (lateral view). cattle. yak.

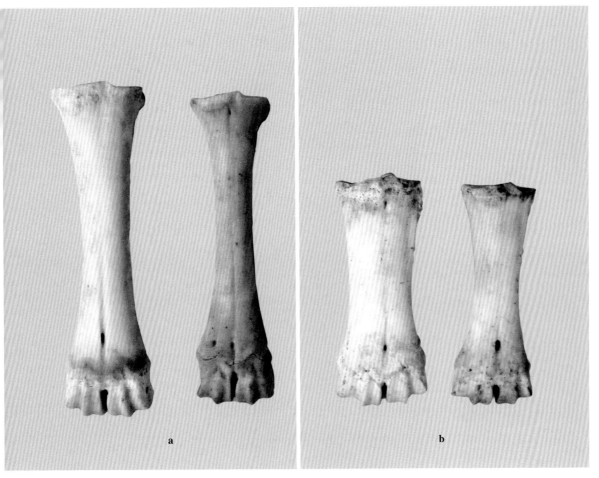

图2 黄牛和牦牛掌骨整体形态（背侧观） a. 黄牛 b. 牦牛
Figure 2. Large metacarpus of cattle and yak (dorsal view). cattle. yak.

图版 27（Plate XXVII）

黄牛与牦牛骨骼形态的对比观察

Differences in Osteological Morphology between Cattle (*Bos taurus*) and Yak (*Bos grunniens*)

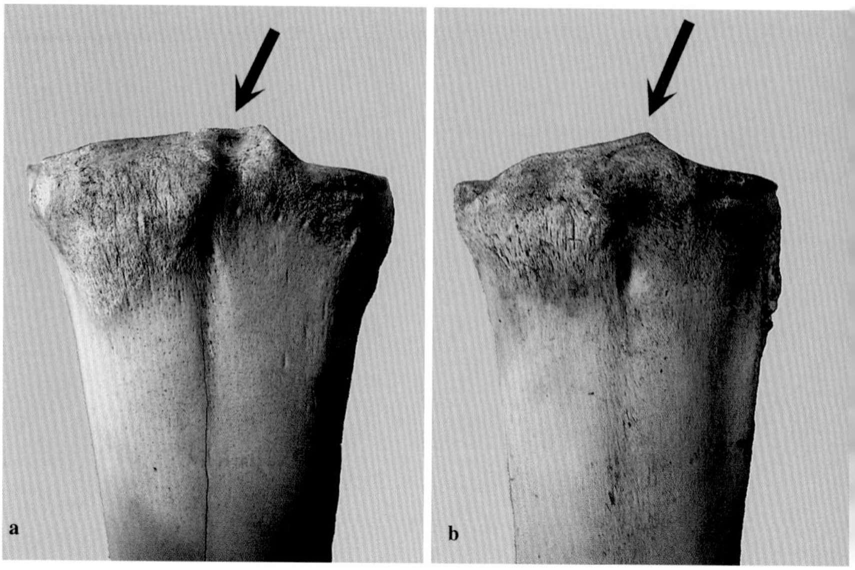

图 1　黄牛和牦牛的掌骨近端（背侧观）

Figure 1. Proximal end of large metacarpus of cattle and yak (dorsal view).

a. 黄牛　b. 牦牛
cattle.　yak.

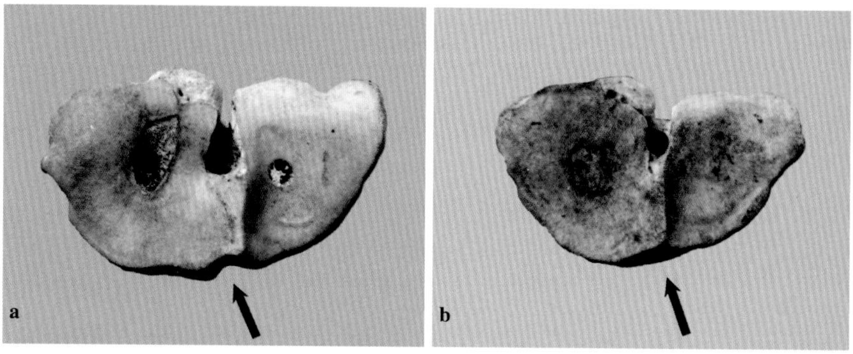

图 2　黄牛和牦牛掌骨近端关节面

Figure 2. Proximal articular surface of large metacarpus of cattle and yak.

a. 黄牛　b. 牦牛
cattle.　yak.

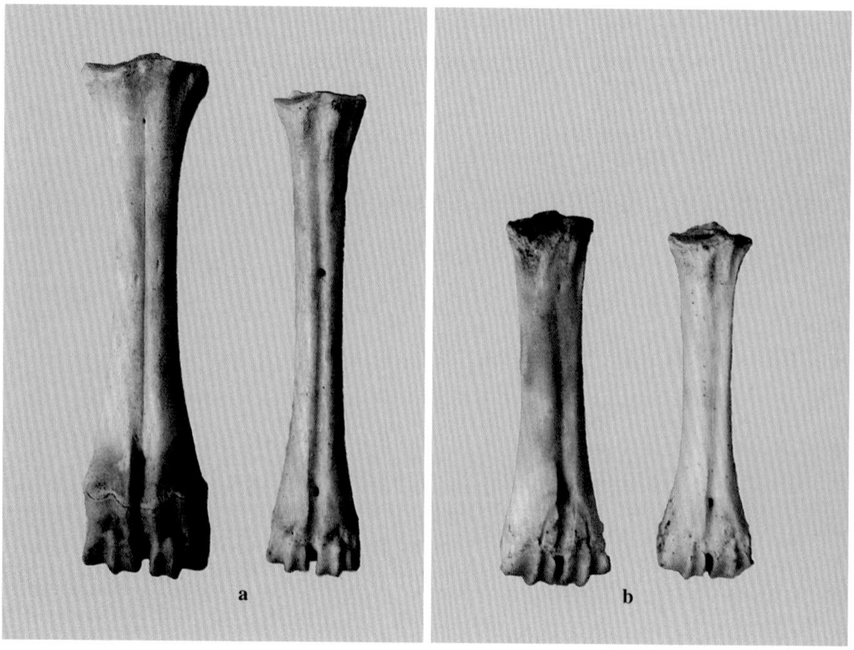

图 3　黄牛和牦牛跖骨整体形态（背侧观）

Figure 3. Large metatarsus of cattle and yak (dorsal view).

a. 黄牛　b. 牦牛
cattle.　yak.

图版 28（Plate XXVIII）

黄牛与牦牛骨骼形态的对比观察
Differences in Osteological Morphology between Cattle (*Bos taurus*) and Yak (*Bos grunniens*)

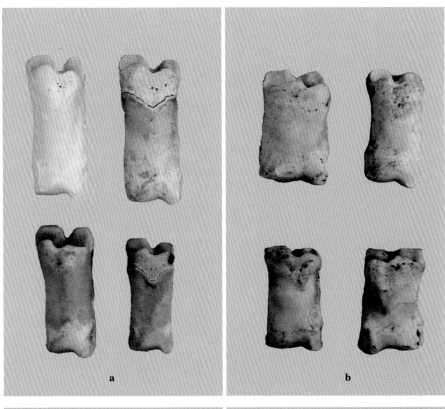

图1 黄牛和牦牛第一指/趾节骨整体形态（背侧观）
Figure 1. First phalanx of cattle and yak (dorsal view).
a. 黄牛　b. 牦牛
　 cattle.　 yak.

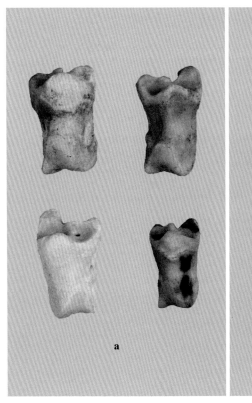
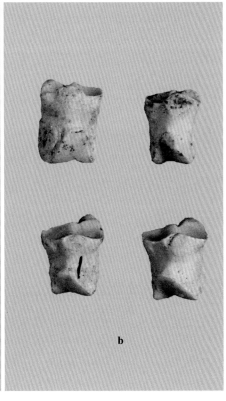

图2 黄牛和牦牛第二指/趾节骨整体形态（背侧观）
Figure 2. Second phalanx of cattle and yak (dorsal view).
a. 黄牛　b. 牦牛
　 cattle.　 yak.

图版29（Plate XXIX）

黄牛与牦牛骨骼形态的对比观察
Differences in Osteological Morphology between Cattle (*Bos taurus*) and Yak (*Bos grunniens*)

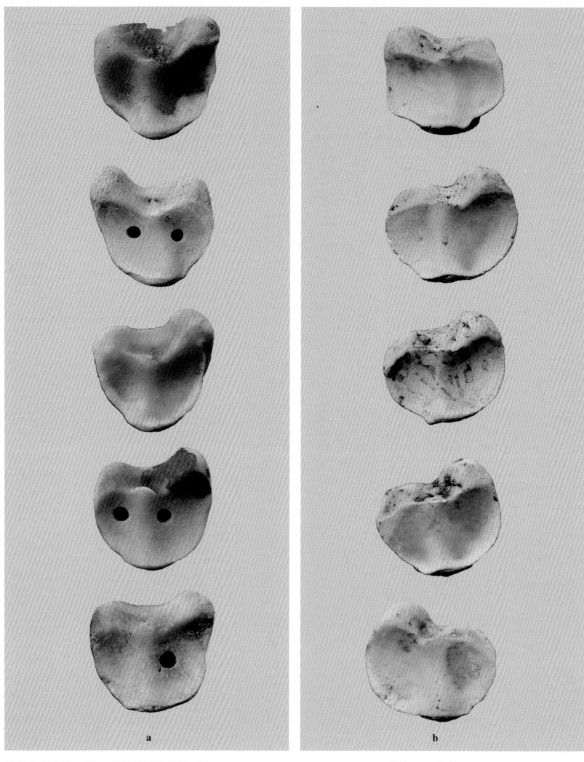

黄牛和牦牛第二指/趾节骨近端关节面形态
Second phalanx of cattle and yak (proximal view).

a. 黄牛   b. 牦牛
   cattle.   yak.

考古遗址出土啮齿目遗存的采集与鉴定方法
Methods of Collection and Identification of Rodent Remains from Archaeological Sites

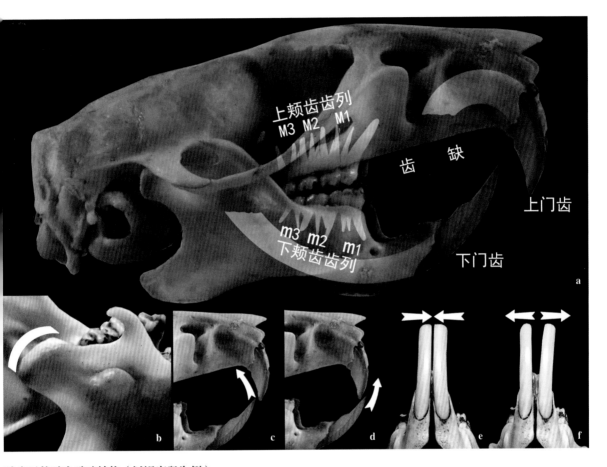

啮齿目的啮合啃咬结构（以褐家鼠为例）
General organization of the rodent dentition (e.g. *Rattus norvegicus*).

a. 门齿和颊齿的相对位置及齿缺
   Relative position of the incisors and cheek teeth, and the diastema in-between.
b. 下颌骨关节突呈狭长的弧背形
   Long and curved condyloid process of the mandible.
c, d. 下门齿可在上门齿的后方与前方做啮合运动
   Lower incisors can occlude posterior or anterior to the upper incisors.
e, f. 下颌联合面松弛，下门齿可做开合运动
   Loose mandibular symphysis, allowing the two lower incisors to move inwards or outwards.

图版 31（Plate XXXI）

考古遗址出土啮齿目遗存的采集与鉴定方法
Methods of Collection and Identification of Rodent Remains from Archaeological Sites

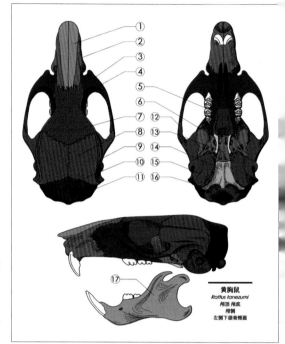

图 1 啮齿目头骨的基本结构
（以黄胸鼠为例，图中编号对应表 1；比例尺为厘米，本文其他图片比例尺均同此）
Figure 1. Bones of the skull in rodents.
(e.g. *Rattus flavipectus*. Numbers correspond with those in Table 1. The scale (black bar) corresponds to a distance of one centimeter.)

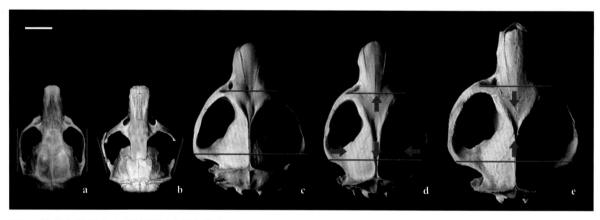

图 2 鼢鼠与竹鼠各自种间的颧弓形态差异
Figure 2. Morphological differences among rodent zygomatic arches. The scale (white bar) corresponds to a distance of one centimeter.

a. 中华鼢鼠的颧弓后部比前部稍宽，或前后部差异不大
For *Myospalax fontanieri*, slight deviation visible between the widths of the anterior and posterior portions of the zygomatic arches.

b. 东北鼢鼠的颧弓前宽后窄
For *Myospalax psilurus*, the anterior portion is broader than the posterior portion of the zygomatic arches.

c. 中华竹鼠的颧弓后部显著扩展
For *Rhizomys sinensis*, the posterior portion of the zygomatic arches is obviously wide.

d. 银星竹鼠的颧弓后部相对较窄，前后相对较长
For *Rhizomys pruinosus*, the posterior portion of the zygomatic arches is relatively narrow and long.

e. 大竹鼠的颧弓前后相对较短，总体更粗壮
For *Rhizomys sumatrensis*, the zygomatic arches are robust, and the distance between the anterior and posterior ends is relatively short.

图版 32（Plate XXXII）

考古遗址出土啮齿目遗存的采集与鉴定方法
Methods of Collection and Identification of Rodent Remains from Archaeological Sites

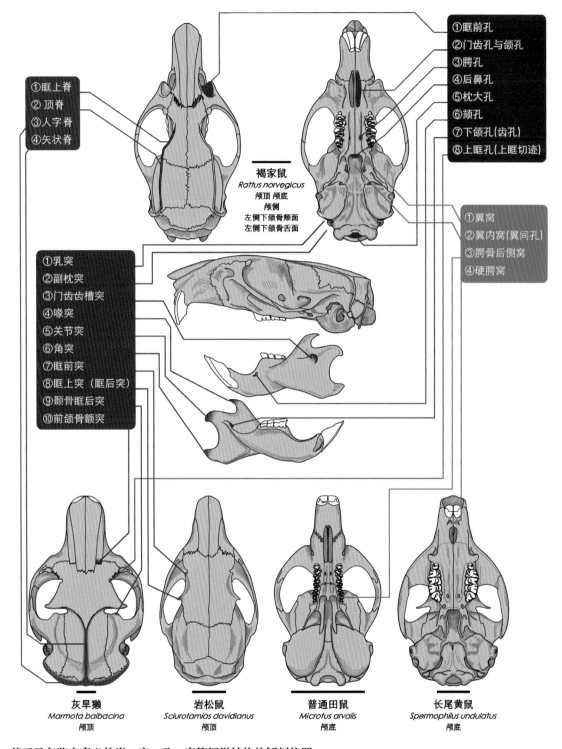

若干具有鉴定意义的脊、突、孔、窝等细微结构的解剖位置
Crests, processes, foramens and foveae of the rodent skulls, which are significant for efficient identification.
The scale (black bar) corresponds to a distance of one centimeter.

图版 33（Plate XXXIII）

考古遗址出土啮齿目遗存的采集与鉴定方法
Methods of Collection and Identification of Rodent Remains from Archaeological Sites

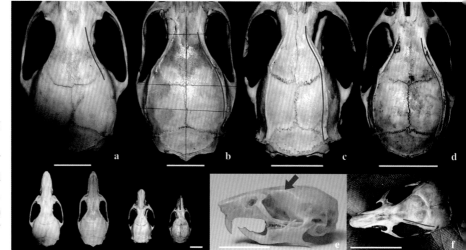

图 1 眶上脊、顶脊的观察比较

Figure 1. Comparison of the supraorbital crests and parietal crests in rodents. The scale (white bar) corresponds to a distance of one centimeter.

a. 青毛鼠的眶上脊较细，长短如红线标示

Supraorbital crest of *Rattus bowersi* is relatively thin and short (marked with red line).

b. 长尾巨鼠的眶上脊更粗更发达，延续更长

Supraorbital crest of *Leopoldamys edwardsi* is developed, much thicker and longer.

c. 褐家鼠的左右顶脊在顶骨位置呈平行发育，与人字脊的左右部分连接

Left and right parietal crests of *Rattus norvegicus* run in parallel, and posteriorly join both ends of the lambdoidal crest.

d. 黑家鼠的眶上脊也较发达，但顶脊呈弧线走向，后端较弱

Supraorbital crest of *Rattus rattus* is developed, while the parietal crest is curved with a weak end.

e. 小林姬鼠眼眶上红色箭头示意位置仅有转折面，并不形成眶上脊

A corner is formed along the parietal bone of *Apodemus sylvaticus* (pointed by red arrowhead) while no supraorbital crest is present.

f. 大林姬鼠眼眶上发育有清晰的眶上脊

Clear supraorbital crests formed on the parietal bones of *Apodemus speciosus*.

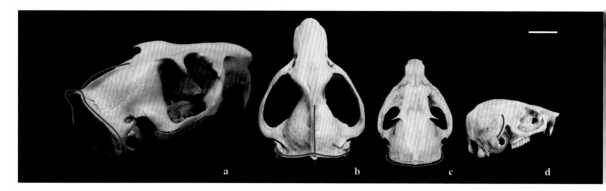

图 2 人字脊、矢状脊的比较

Figure 2. Comparison of the lambdoidal crests and sagittal crests in rodents. The scale (white bar) corresponds to a distance of one centimeter.

a, b. 喜马拉雅旱獭和中华竹鼠均具发达的人字脊和矢状脊

Both *Marmota himalayana* and *Rhizomys sinensis* have developed, robust lambdoidal crests and sagittal crests.

c. 长尾黄鼠只有发达的人字脊，无矢状脊

Spermophilus undulatus only has developed lambdoidal crest, but no sagittal crest.

d. 中华鼢鼠的人字脊的左右部分发达，如同听泡的上檐

For *Myospalax fontanieri*, the left and right portions of the lambdoidal crest are well developed, forming the covering structure for the auditory vesicles.

考古遗址出土啮齿目遗存的采集与鉴定方法
Methods of Collection and Identification of Rodent Remains from Archaeological Sites

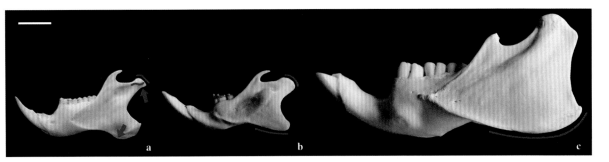

图 1　下颌骨乳突、关节突、角突的比较

Figure 1. Comparison of the angular process, condyloid process and coronoid process in rodent mandibles. The scale (white bar) corresponds to a distance of one centimeter.

a. 松鼠型的关节突显得方正、带颈、有结节，角突向下方突起
Condyloid process of sciuromorphs looks relatively square and has a small tubercle. The angular process makes a forward extension, forming a so called vice angular process.

b. 鼠型的关节突显得更接近圆弧，角突不向下方突起
Condyloid process of myomorphs is roundish and blunter. The angular process shows a weak protrusion.

c. 豪猪型的喙突明显退化，有的接近消失，关节突发达，角突下方呈圆弧形
Coronoid process of hystricomorphs is degenerated, in some cases even nearly absent. The condyloid process is especially robust and the lower part of the angular process is arc-shaped.

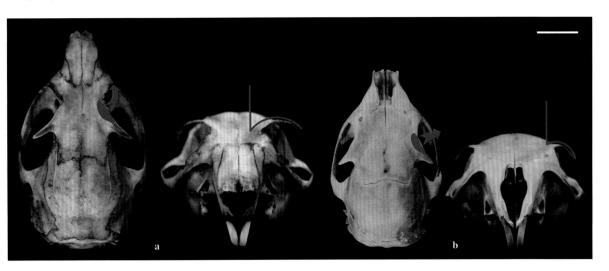

图 2　鼯鼠族与松鼠族的眶上突差异

Figure 2. Morphological differences of the supraorbital process between Pteromyini and Sciurini. The scale (white bar) corresponds to a distance of one centimeter.

a. 鼯鼠族（图中以复齿鼯鼠为例）的眶上突、红色块面所示意部位，往上翻翘，眶上突尖尾往上展起的角度也更大
Supraorbital process (marked in red) of Pteromyini (e.g. *Trogopterus xanthipes*) tilts upwards, and posterior end stretches at a relatively wide angle.

b. 松鼠族（以蓝腹松鼠为例）的眶上突、红色块面所示意部位，仅往外水平延展，并不向上翘起，眶上突尖尾往上展开角度小
Supraorbital process (marked by red) of Sciurini (e.g. *Callosciurus pygerythrus*) extends out horizontally, and the posterior end stretches at a relatively narrow angle.

考古遗址出土啮齿目遗存的采集与鉴定方法
Methods of Collection and Identification of Rodent Remains from Archaeological Sites

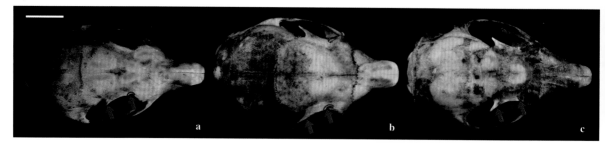

图 1　岩松鼠、红松鼠与花鼠的眶上突及眶前突

Figure 1. Supraorbital process and preorbital process of *Sciurotamias davidianus*, *Sciurus vulgaris* and *Tamias sibiricus*. The scale (white bar) corresponds to a distance of one centimeter.

a. 岩松鼠的眶上突和眶前突均又薄又小

　Both the supraorbital and preorbital processes of *Sciurotamias davidianus* are thin and small.

b. 红松鼠的眶上突较发达，有眶前突

　Sciurus vulgaris displays relatively developed supraorbital and preorbital processes.

c. 花鼠的眶上突中度发达，有眶前突

　Tamias sibiricus exhibits moderately developed supraorbital processes, and has small preorbital processes.

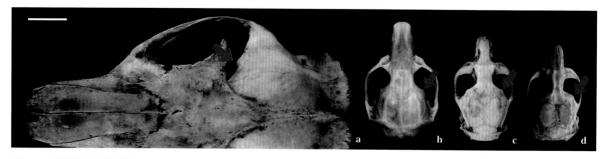

图 2　颞骨眶后突的比较

Figure 2. Comparison of the postorbital process on the temporal bone. The scale (white bar) corresponds to a distance of one centimeter.

a. 灰旱獭具有其他旱獭没有的颞骨眶后突

　Marmota baibacina shows the postorbital process on the temporal bone which is absent among other species of marmots.

b–d. 中华鼢鼠、蒙古田鼠以及社田鼠均有显著的颞骨眶后突

　Myospalax fontanieri, *Microtus mongolicus* and *Microtus socialis* all have remarkable postorbital processes on the temporal bone.

图版 36（Plate XXXVI）

考古遗址出土啮齿目遗存的采集与鉴定方法
Methods of Collection and Identification of Rodent Remains from Archaeological Sites

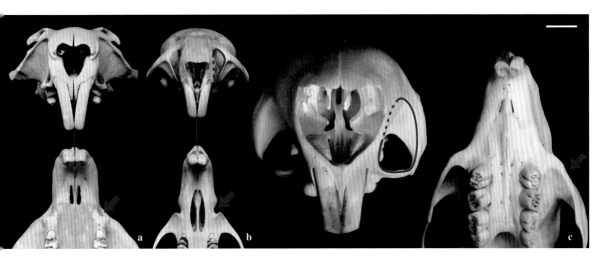

图 1 松鼠型、鼠型和豪猪型啮齿目的眶前孔

Figure 1. Antorbital foramens in sciuromorphs, myomorphs, and hystricomorphs. The scale (white bar) corresponds to a distance of one centimeter.

a. 松鼠型（以中华旱獭为例）的眶前孔是一个很小的穿孔，孔的下方腭板上有显著突起，用于附着咬肌

 Antorbital foramen of sciuromorphs (e.g. *Marmota Himalayana*) is a small hole, underneath which a process on the plate is formed to attach the masseter muscle.

b. 鼠型（以青毛鼠为例）的眶前孔较大

 Antorbital foramen of myomorphs (e.g. *Rattus bowersi*) is relatively large.

c. 豪猪型（以中华豪猪为例）的眶前孔几乎与眼眶等大

 Antorbital foramen of hystricomorphs (e.g. *Hystris hadgsoni*) is almost the same size as the orbit.

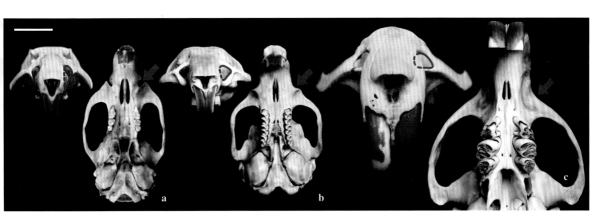

图 2 鼠型啮齿目的三种典型眶前孔

Figure 2. Three types of antorbital foramens in myomorphs. The scale (white bar) corresponds to a distance of one centimeter.

a. 鼠科的（以褐家鼠为例）钥匙孔型

 Key-shaped type of Muridae (e.g. *Rattus norvegicus*).

b. 鼢鼠科的（以中华鼢鼠为例）竖三角型

 Vertical triangular type of Myospalacidae (e.g. *Myospalax fontanieri*).

c. 竹鼠科（以中华竹鼠为例）的横三角型

 Horizontal triangular type of Rhizomvidae (e.g. *Rhizomys sinensis*).

考古遗址出土啮齿目遗存的采集与鉴定方法
Methods of Collection and Identification of Rodent Remains from Archaeological Sites

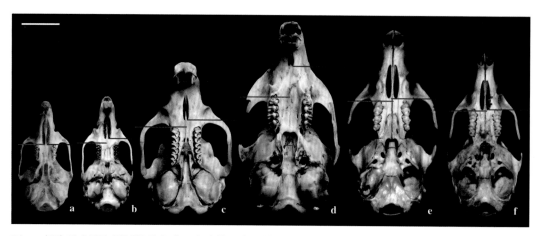

图 1　门齿孔末端和最近的前臼齿 / 臼齿的距离比较
Figure 1. Distances between posterior end of the Stensen's foramen and cheek teeth in some rodents. The scale (white bar) corresponds to a distance of one centimeter.

a. 小林姬鼠门齿孔末端与 M1 齐平
   Posterior end of the Stensen's foramen in *Apodemus sylvaticus* is close to M1.
b. 大林姬鼠的则具有显著距离
   Posterior end of the Stensen's foramens of *Apodemus speciosus* is relatively far from M1.
c. 中华鼢鼠的门齿孔窄而短，距离 M1 较远
   Stensen's foramen of *Myospalax fontanieri* is narrow and short, clearly far from M1.
d. 岩松鼠的门齿孔仅位于前颌骨，前端靠近门齿，末端距离第一枚前臼齿 P3 很远
   Stensen's foramen of *Sciurotamias davidianus* only locates on the premaxilla, close to the incisors while remarkably far from P3.
e, f. 褐家鼠和黑家鼠的门齿孔均窄而长，抵近臼齿 M1
   Rattus norvegicus and *Rattus rattus* both have long and narrow Stensen's foramens, close to first upper molars.

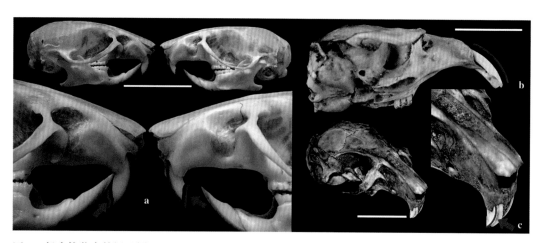

图 2　门齿的鉴定特征示例
Figure 2. Identifying features of incisors in rodents. The scale (white bar) corresponds to a distance of one centimeter.

a. 左侧小家鼠的上门齿齿尖后缘有标志性缺刻，右侧小林姬鼠的则没有
   Mus musculus (left) bears a distinctive notch on the wear facet of the incisor, while *Apodemus sylvaticus* (right) does not show this feature.
b. 鼹形鼠的门齿向前显著伸突
   Incisors in *Ellobius* sp. protrude remarkably.
c. 长爪沙鼠上门齿的前表面有一条纵沟
   Long grooves are formed on the anterior surface of the incisors in *Meriones Unguiculatus*.

大数据及其在动物考古学中的应用
The Application of Metadata in Zooarchaeology

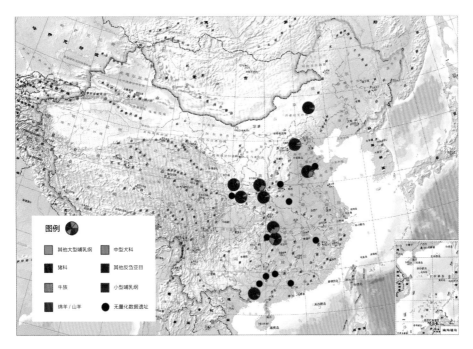

图 1　新石器时代早期各类哺乳动物骨骼相对比例
Figure 1. Relative proportion of mammal categories of the Early Neolithic period.

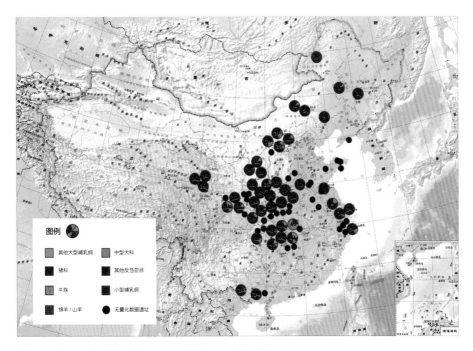

图 2　新石器时代中期各类哺乳动物骨骼相对比例
Figure 2. Relative proportion of mammal categories of the Middle Neolithic period.

图版 39（Plate XXXIX）

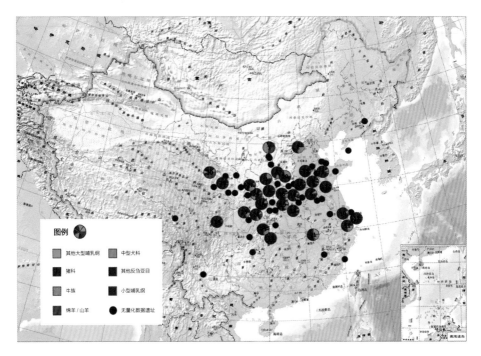

图 1　新石器时代晚期各类哺乳动物骨骼相对比例

Figure 1. Relative proportion of mammal categories of the Late Neolithic period.

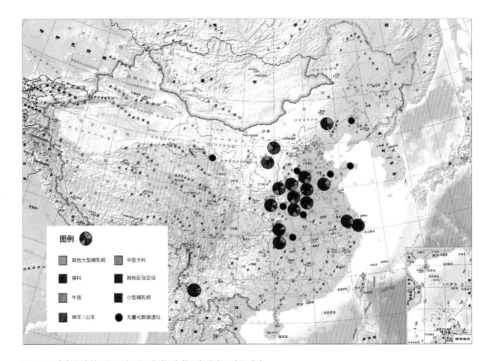

图 2　青铜时代早期各类哺乳动物骨骼相对比例

Figure 2. Relative proportion of mammal categories of the Early Bronze Age.